Optical instruments
and
their applications

By the same author

Optical production technology
Dividing, ruling and mask-making
Lens mechanism technology
Spectacle lens technology

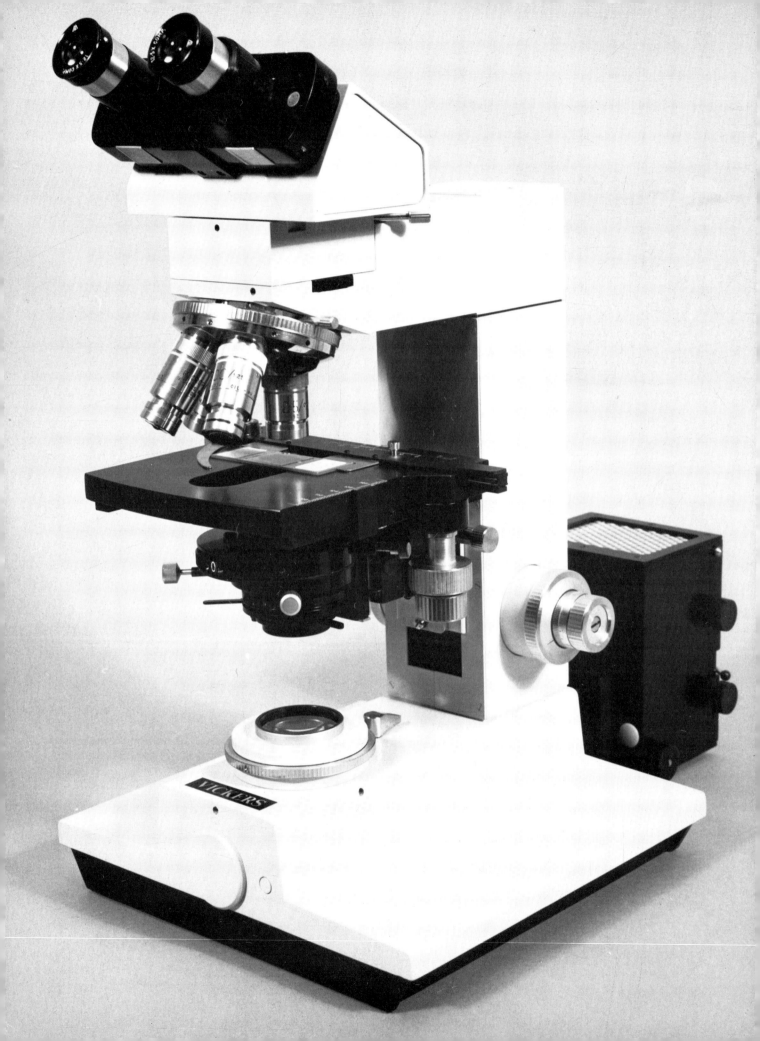

Optical instruments and their applications

Douglas F. Horne, M.B.E.

D.Sc., C.Eng., F.R.Ae.S., F.I.Prod.E.

Senior Lecturer, School of Production Studies, Cranfield Institute of Technology
Group Production Adviser, Hilger and Watts Ltd, 1962–69
General Manager (G. B.–Kalee Division), Rank Precision Industries Ltd, 1958–62
Group Production Engineer, Rank Precision Industries Ltd, 1948–58

Adam Hilger Ltd, Bristol

British Library Cataloguing in Publication Data
Horne, Douglas Favel
Optical Instruments and their Applications
1. Optical instruments
I. Title
681′.4   TS513
ISBN 0–85274–345–9

First published 1980

Published by Adam Hilger Ltd,
Techno House, Redcliffe Way,
Bristol, BS1 6NX.
The Adam Hilger book-publishing imprint is
owned by The Institute of Physics.

Typeset by Preface Ltd, Salisbury and
printed and bound in Great Britain by
William Clowes (Beccles) Limited
Beccles and London

To my wife

Author's Preface

When the three books, *Optical production technology*, *Dividing, ruling and mask-making* and *Lens mechanism technology*, had been published, I realised that there was no book available for a design engineer or instrument maker interested in the manufacture of equipment containing optical assemblies. Optical elements are often at the very heart of a system that may control a machine, an industrial process, or the flight of an aircraft.

How would you survey a country, construct a block of flats or build a ship without a telescope? Would modern society be possible without cameras and projectors for making motion pictures, television, illustrated books, maps and journals? How can scientists, metallurgists and biologists see the smallest particles of matter without a microscope? Industrial progress is based on precision measurement using optical scales, graticules and gratings. Quality control in the production of steel, non-ferrous materials, chemicals and gases is achieved by spectrochemical analysis.

I have endeavoured to classify different types of instrument in separate chapters, but theodolites include both telescopes and reading microscopes. Document-copying machines, engineering projectors, microfilm readers, phototypesetters and colour scanners for printing newspapers, journals and books can project images onto photosensitive material, and therefore may be classified as cameras.

The importance of the optical industry is not always obvious to those outside it, and the object of this book is to describe typical instruments and their applications with the hope that able young engineers will want to find a career in this fascinating industry.

The ingenuity of engineering designers is essential for economic production, and since 1950 the electronic digital computer has helped to improve the performance of optical instrumemts and has challenged the initiative of production engineers responsible for their manufacture.

Superior designs of optical equipment ensure the confidence of marketing executives who will enthusiastically forecast large-volume sales. Expanding production programmes justify capital expenditure on new buildings, modern plant and machinery that result in high productivity. Cash flow and profit from satisfied customers make possible the development of new designs.

Our rapidly expanding microelectronics industry will produce, in the 1980s, many sophisticated devices that have applications in optical instruments. In particular, the charge-coupled device, a form of metal-oxide semiconductor microcircuit, has the capability of being used as an imaging device of high resolution suitable for solid-state colour-television cameras. These developments will influence the design of photographic equipment and, gradually, the silver-emulsion-coated film, used in motion pictures, will be replaced by low-cost electronic video recording with compact storage on magnetic tape or in random-access solid-state memories.

During more than 20 years in the optical industry at group production engineering level, I have had experience with the manufacture of many types of instrument similar to those described and illustrated in this book. It is unfortunate that company reorganisation from a group structure, based on the development and manufacture of a wide range of products, to a division structure, giving priority to marketing a narrow range of products, has now made it very difficult for any one person to gain a broad knowledge of the optical industry. Specialisation has long-term disadvantages for technologists and engineers.

The British optical industry has developed over more than 250 years, from the time of Sir Isaac Newton (1643–1727), and it is impossible to mention all those who have contributed to progress in this period. The first chapter, which is concerned with the history of the British optical industry, has therefore been limited to companies currently manufacturing optical equipment and to some of the personalities who were responsible for design or production. Companies are entered in alphabetical order of names by which they were known in 1970. This historical information should interest readers in countries where an optical industry has developed during the twentieth century.

Over 300 photographs and more than 160 charts and diagrams illustrate the optical instruments described herein, instruments that are manufactured by leading companies in Great Britain, Europe and America. Without their cooperation, this book would not have been possible. I hope that I have not made any technical errors or significant omissions in the text, but, if I have, I would like to be advised, so that corrections or additions can be made in the future.

Any opinions that I have expressed are, of course, my responsibility and do not indicate the policy of the Cranfield Institute of Technology, The Institute of Physics, Adam Hilger Ltd or any company that has given information to me.

Permission has been given by the following companies or publishers for reproduction of photographs and technical information. The names are listed in the order in which material appears for the first time in the book.

Barr and Stroud Ltd (Mr T. Johnston)
Chance–Pilkington (Mr A. B. Scrivener)
J. H. Dallmeyer, Ltd
Dollond and Aitchison Group Ltd (Mr C. Howell)
Ealing Beck (Mr A. K. Howgego)
Grubb–Parsons (Mr G. E. Manville)
Wray (Optical Works) Ltd (Mr A. Smith)
W. Ottway and Co. Ltd (Mr G. C. Ottway)
United Kingdom Optical (Mr A. A. S. Moore)
Univis Lens Company (Mr S. A. Emerson)
Vickers Instruments (Mr A. J. Munro)
V.E.B. Carl Zeiss Jena (Professor Dr Mutze)
Royal Observatory, Edinburgh (Mr G. J. Carpenter)
Mullard Ltd (Mrs A. Healy)
American Optical Corporation (Mr A. J. Liberty)
Wild Heerbrugg AG (Mr W. Piske)
Arnold and Richter KG (Mr R. Schutz)
Linhof Prazisions Kamera Werke GmbH (Dr Pientka)
Rank Xerox Ltd (Mr D. J. Day)
Pictorial Machinery Ltd
Littlejohn Graphic Systems Ltd (Mr A. R. G. Stephenson)
The Monotype Corporation Ltd (Mr M. B. Davies)
Linotype U.K. (Mr A. Bluhm)
Linotype-Paul (Mr F. Cannings)
Crosfield Electronics (Mr P. C. Pugsley)
Muirhead Data Communications Ltd (Mr R. A. Everett)
Buckbee–Mears Europe GmbH (Mr H. Neupert)
RCA Systems Division (Mr D. S. Clarke)
Rank Taylor Hobson (Mr M. James)
Kern and Co. AG (Mr A. Kunzli)
Carl Zeiss (Mr R. Eilenberger)
Hunting Surveys and Consultants Ltd (Mrs M. van Dijk)
Matra Division Optique (Mr J de Montremy)

Henri Hauser Ltd, Bienne
Hewlett Packard (Mr T. M. Hoffer)
Acuity Systems Inc. (Mr D. Voracek)

I also wish to thank my wife for typing the manuscript and our family for their help during the time of writing this book. The care and interest taken by the staff of Adam Hilger Ltd whilst editing this book is much appreciated.

March 1980   D. F. Horne

Contents

1 History of the British Optical Industry, 1720–1970

1.1 Barr and Stroud

Rangefinders

On 25 May 1888, the Admiralty advertised in *Engineering* requesting proposals for a rangefinder capable of being carried by one infantry soldier and able to find the range of fixed objects at a distance of 2500 yd (2100 m) (Fig. 1.1.1).

Dr Archibald Barr, Professor of Engineering, and Dr William Stroud, Professor of Physics, both at Yorkshire College (now the University of Leeds), within two weeks developed the general construction and principles for a short-base rangefinder. A patent application was lodged on 30 June 1888 and accepted on 29 June 1889 (Fig. 1.1.2).

Among the inventions covered by the patent were three basic ideas which distinguished the Barr and Stroud rangefinder from those of other designs.

(*a*) *Coincidence method of setting*. The two images, formed by light entering the right and left ends of the instrument, were made to give entirely separate images, one above and one below a line of separation. The angle subtended by a target, at the base of a rangefinder, caused the two images to be displaced horizontally and the amount of displacement was proportional to the range. This technique, capable of high accuracy, was explained by the

TENDERS.

RANGE FINDERS FOR INFANTRY, FOR H.M. SERVICE.

HIS ROYAL HIGHNESS the COMMANDER-IN-CHIEF is prepared to receive

Proposals from Persons desirous of submitting for competitive trial RANGE FINDERS, which shall fulfil the following conditions, viz. :—

(1). The instrument must be strong and simple and must not require frequent adjustment, nor extraordinary precautions against the exigencies of Field Service and effects of climate and weather.

(2). It must be possible for one Infantry Soldier to carry the entire apparatus, in addition to his rifle and ordinary equipment, both on the march and in action.

(3). A person of ordinary intelligence and normal vision must be able to become an efficient range taker after a month's training.

(4). The system must be suitable for the observation of moving objects, especially troops of all arms in the usual formations up to 1200 yards distance.

(5). It must be practicable to make at least four observations per minute with an average error, at 1000 yards, not exceeding 4 per cent. of the range.

(6). The range must be read in yards without recourse to calculation.

(7). The number of observers required for each instrument or set of instruments must not exceed two.

(8). Other considerations being equal, preference will be given to those instruments which

(*a*) Can find the range of fixed objects up to 2500 yards with the greatest rapidity, but with an error not exceeding 100 yards; this to be effected either by the normal mode of operations, or with the aid of extra appliances.

(*b*) Require only one observer.

(9). The instruments submitted for trial must be complete, with full description, and instructions for use.

(10). Proposals must be sent in, addressed to the ADJUTANT-GENERAL, Horse Guards, War Office, before the 1st August. 1888. Q 306.

1.1.1

Advertisement in *Engineering* on 25 May 1888 by the War Office for a rangefinder.

1.1.2
Early rangefinder supplied against the first Admiralty order.

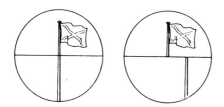

1.1.3
The coincidence method of setting.

professors during a lecture at the Institution of Mechanical Engineers in January 1896. It is now used in many types of camera viewfinder (Fig. 1.1.3).

(*b*) *The translating deflecting prism.* The displacement of the partial images is measured by bringing them into coincidence by means of a refracting prism P of small angle, placed in the converging light beam, and able to move longitudinally along the path of the rays. When the prism is moved from P to P′, the position of the image changes from C to C′, so the longitudinal motion of the prism is a measure of the displacement of the partial images (Fig. 1.1.4).

In the early rangefinders, the prism moved 6 in to measure the image displacement between infinity and 250 yd (210 m) range. The motion of a prism required to measure an angle of one second of arc is 1/200 in. The separation of two images, in the field of view, for an angle of one second is 1/8000 in, so the translating prism gave a 40× magnification.

(*c*) *Range scale attached to the translating prism.* The attachment of a scale to the moving prism eliminated any possibility of error due to imperfections in the mechanism. This feature is one of the reasons for the superiority of Barr and Stroud instruments.

The first instrument manufactured by Barr and Stroud for use in the Army was not accepted, but in 1891 they were asked to design a naval rangefinder, which was fitted to H.M.S. Arethusa at Chatham in 1892, and during 1893 five instruments were ordered. All the important optical components, including the prisms, were made for them by Adam Hilger Ltd.

By 1895 there was a considerable demand for naval rangefinders from many foreign countries, except Germany, and this was followed by the design and construction of fire control instruments. By 1904, in a new factory at Anniesland, Glasgow, optical production was established and expanded until 1912 when, after 24 years partnership, the firm of Barr and Stroud (then employing about 1000 people) was made into a Limited Company with Dr A.

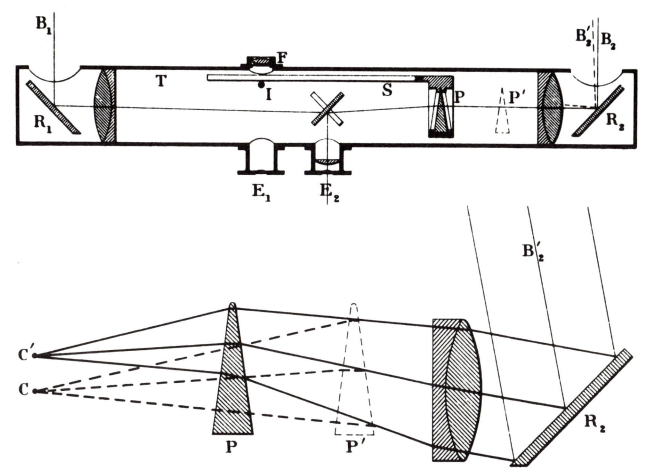

1.1.4
The translating deflecting prism P for rangefinders. Later designs used pentagonal prisms instead of the mirrors R_1 and R_2.

Barr as Chairman. In 1917, Francis Morrison and John Martin Strang were appointed additional Directors of Barr and Stroud Ltd.

There was a very great improvement in accuracy of range measurement over this period of development. The original instrument of 1888 had an error of 4% or 40 yd in 1000 yd. The 1892 naval rangefinder had improved to 3% at 3000 yd, which was the effective range of guns at that time. By 1916, large naval guns had a range of 25000 yd and the rangefinders an error of less than 250 yd.

Periscope rangefinders

In 1916, Barr and Stroud Ltd were given their first order for six submarine periscope rangefinders. The periscope had an optical length of 30 ft with a 3 ft rangefinder base length. At that time, the main periscope tube was 6 in (150 mm) diameter and of monocular design, which was a cause of eye strain with submarine officers who had long periods of observation duty. Dr Stroud solved the problem of periscopic rangefinding and, in 1917, 20 pairs of complete periscopes, including rangefinders, were ordered by the Admiralty.

The following innovations were embodied in these early Barr and Stroud periscopes.

(*a*) *Internal focusing.* Previous designs of periscope had separate focusing lenses, on the outside of the eyepiece, to suit the observer's eyes. Barr and Stroud arranged for one of the internal lenses of the periscope to be moved by a knurled head and so give a continuous adjustment.

(b) *Internal colour glasses*. Instead of fitting an external cap over the eyepiece, the colour glasses were built into the optical system.

(c) *Control handles*. The handles for moving the periscope in azimuth were provided with rotating grips for the control of magnification and movement in elevation of the top prism.

(d) *Range estimator*. A device was fitted in the periscope to enable the range and inclination of the target to be found. The angle subtended by the height of the target was measured and so, with a known target height, the range could be easily calculated.

Binocular periscopes

In 1924, after months of discussion with Barr and Stroud, the Admiralty ordered their first binocular periscope which was to be 40 ft long and 9.5 in diameter. The design embodied lenses and prisms which enabled the light from two windows to be passed down the main tube through lenses common to both beams. The lenses 1 and 2 were wedge-shaped (Fig. 1.1.5(a)) so that they diverted the light beams towards the common lenses 3 and 4. The bottom prism 5 was designed so that the beams were crossed, and light reaching the right eyepiece entered from the right side of the top prism 1. In this way, true stereoscopic views were presented to the observer, without eye strain, and permitted distant objects to be examined and identified. The binocular periscope was therefore adopted as the 'look-out' or 'cruising' periscope (Fig. 1.1.5(b)).

Development of Barr and Stroud Monocular and Binocular Periscopes has continued to permit higher underwater speeds and increased periscope depth. With the introduction of 'snort' masts, better navigation and watch-keeping facilities were needed, so anti-vibration optical systems, artificial horizons, radar ranging and astronomical sextants were fitted to enable satisfactory control at submerged speeds well over 10 knots.

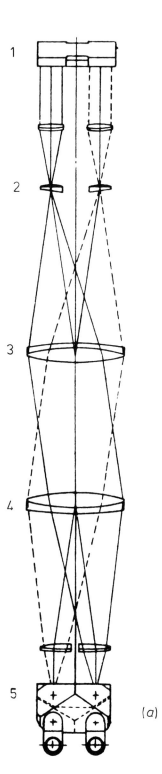

(a)

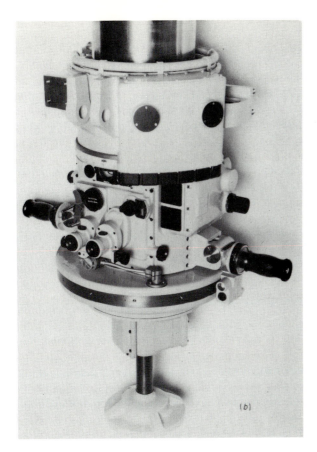

1.1.5

(a) Opposite: Diagrammatic view of optical system for submarine binocular telescope Type CK1. (b) Above: Lower end of a binocular submarine periscope.

In the 1950s and 1960s, there have been many applications for remote-viewing optical systems required in nuclear reactor, chemical and general engineering industries. A special periscope, 100 ft long, was produced for the House of Commons so that the control engineer can see how many Members of Parliament and 'Strangers' are present. From visual observations, adjustments are made to the ventilation system to ensure comfortable conditions.

Binoculars

Galilean binoculars, in the form of opera glasses, and prismatic binoculars, for use on ships' bridges, have been manufactured by Barr and Stroud since 1919, when the first night binocular was sent for competitive trials. More than 10 years later, in December 1930, their first order for prismatic night binoculars was received from the Admiralty (Fig. 1.1.6).

The light-gathering power of a binocular depends on the aperture of its objectives, and a nominal 50 mm diameter is as large as can be efficiently used for night glasses. The exit pupil of 7 mm diameter corresponds to the maximum size of the pupil of a human eye in the dark, so the observer's eyes are flooded with light under all conditions of illumination. Any increase in the aperture of the objective does not give an increase in the amount of light reaching the eye as the exit pupil would be greater than the pupil of the eye.

Optical reflector gunsights

A pilot's sight presented a design problem, as the pilot must have considerable freedom of eye position without distortion of the sighting mark during use. This required a very large relative aperture in the optical system which projected the mark into the line of vision. The first design achieved this by using a paraboloidal metal reflector, and type GD5, covered by a patent of 1934, was made in considerable quantities.

In 1937, an improved design, substituting a lens system for a mirror, with a numerical aperture of $f/0.68$, was made in very large quantities before and during the 1939–45 Second World War. Type GM2 sight (Fig. 1.1.7) had the following functions:

(a) To give the line of sight to a target.

(b) To determine when to fire on a target at a previously determined range.

(c) To estimate the range of a target at any instant.

The sectional view of the sight shows that the objective of 3·6 in aperture consists of four lenses. The graticule, illuminated by a lamp, is projected through the objective onto a screen and reflected to the pilot's eye. As the reflector

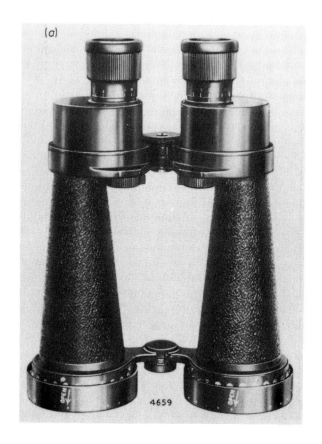

(a)

4659

is made of clear glass, the target can be simultaneously viewed through it. The spot (S) in the centre of the graticule and the gap in the setting of adjustable horizontal rods (R) give the means of measuring the range, providing the size of the target is known. The range scale is graduated from 150 to 600 yd and the base scale, represented by the gap between the rods, from 40 to 100 ft.

1.1.6
(a) A 7× 50 mm prismatic binocular with ray shade. **(b)** Prismatic night binocular manufactured by Barr and Stroud.

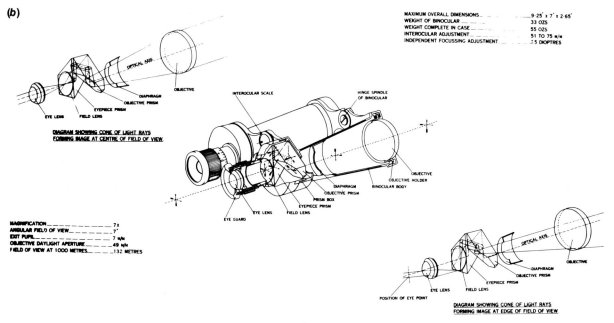

(b)

MAXIMUM OVERALL DIMENSIONS	9·25˝ x 7˝ x 2·65˝
WEIGHT OF BINOCULAR	33 OZS
WEIGHT COMPLETE IN CASE	55 OZS
INTEROCULAR ADJUSTMENT	51 TO 75 M/M
INDEPENDENT FOCUSSING ADJUSTMENT	± 5 DIOPTRES

DIAGRAM SHOWING CONE OF LIGHT RAYS FORMING IMAGE AT CENTRE OF FIELD OF VIEW.

OPTICAL AXIS
OBJECTIVE
DIAPHRAGM
OBJECTIVE PRISM
EYEPIECE PRISM
FIELD LENS
EYE LENS

INTEROCULAR SCALE
HINGE SPINDLE OF BINOCULAR
OBJECTIVE
OBJECTIVE HOLDER
BINOCULAR BODY
DIAPHRAGM
OBJECTIVE PRISM
PRISM BOX
EYEPIECE PRISM
FIELD LENS
EYE LENS
EYE GUARD

MAGNIFICATION	7x
ANGULAR FIELD OF VIEW	7°
EXIT PUPIL	7 M/M
OBJECTIVE DAYLIGHT APERTURE	49 M/M
FIELD OF VIEW AT 1000 METRES	132 METRES

OPTICAL AXIS
OBJECTIVE
DIAPHRAGM
OBJECTIVE PRISM
EYEPIECE PRISM
FIELD LENS
EYE LENS
POSITION OF EYE POINT

DIAGRAM SHOWING CONE OF LIGHT RAYS FORMING IMAGE AT EDGE OF FIELD OF VIEW.

6

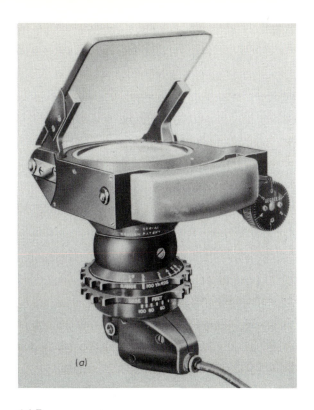

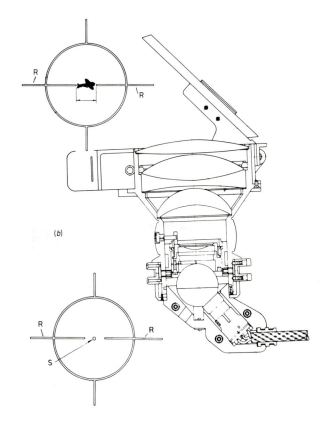

1.1.7
(a) Type GM2 Barr and Stroud optical reflector gunsight and range estimator for aircraft. (b) Schematic diagram.

Air survey and photogrammetry

During 1924, the Air Survey Committee of the War Office, established in 1920 to initiate research in apparatus and methods, approached Barr and Stroud Ltd to see if they would be willing to design and manufacture instruments for the new technique of air survey.

The first instrument was a Topographical Stereoscope for use in the field or at Army establishments to enable stereoscopic observation to be made from pairs of aerial photographs (Fig. 1.1.8). The two photographs to be viewed were placed under grid plates and, when seen stereoscopically, the grid lines had the appearance of a net stretched over the

ground. The grid plates were adjustable so that they could be brought nearer to each other or moved further apart. The effect of this movement was to raise or lower the net so that it appeared to be level with any chosen feature, and any ground features higher or lower could be observed. This simple instrument, delivered in 1927, provided a method for the rapid drawing of small-scale maps.

A much better orientation of the photographs could be obtained with the Precision Stereoscope Type ZD15 which had glass grids, with fine etched diagonal lines, instead of the topographical stereoscope grid. The grid base lines were in the same straight line and each photograph was mounted on a turntable with its centre on the grid base line (Fig. 1.1.9).

At this time, Lieutenant M. Hotine, R.E., later to become Major General in charge of the Ordnance Survey at Southampton, was a frequent visitor to Glasgow and a friend of Dr A. Barr.

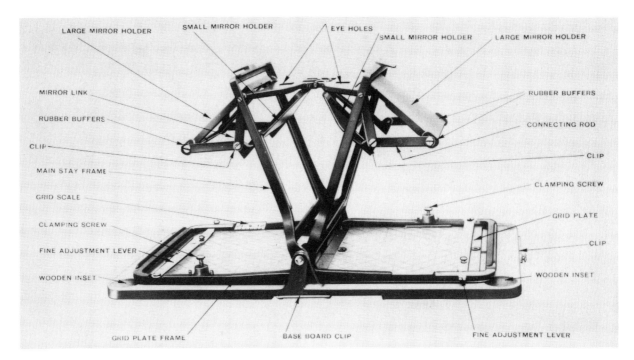

1.1.8
Barr and Stroud ZD4 Topographical Stereoscope for examination of aerial photographs, with 60% overlap, covered by a parallactic grid.

1.1.9
Barr and Stroud Precision Topographical Stereoscope ZD15.

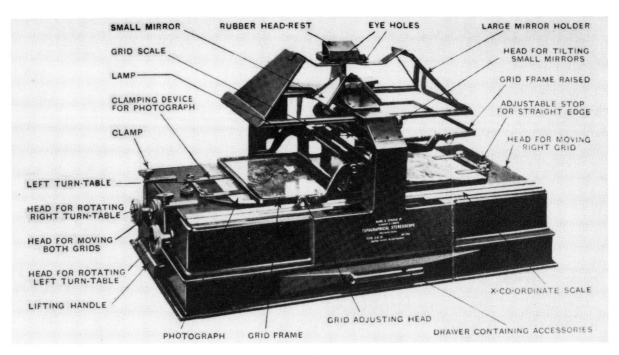

1.1.10
**Barr and Stroud Epidiascope for revising
maps from aerial survey photographs.**

A rapid and convenient means for revising old maps, by the insertion of new details, was needed, so an epidiascope was designed, which is similar to the projector epidiascope used for illustrating lectures. A photograph was placed on the copy board and a semi-transparent impression from the map was placed on the screen. The scale of the two pictures was adjusted along the optical axis and the copy board was given a tilt to equal that at the time of exposure of the aerial photograph. Adjustments were made by trial and error until the projected images of the control points coincided, and then the detail could be traced onto the map (Fig. 1.1.10).

In 1925, an order was placed with Barr and Stroud Ltd for a Photogrammetric Plotting Machine to make large-scale maps working on either stereoscopic or coincidence principles. The plotter was designed by Dr A. Barr to produce, on the drawing board, a map of portions of the ground, represented by two aerial photographs, and to draw contour lines on the map (Fig. 1.1.11).

Based on the patents of Mr H. G. Fourcade, of South Africa, two other types of photogrammetric plotters were ordered from Barr and Stroud Ltd for design and construction. These were the Fourcade Stereogoniometer and the Thompson Plotter, Lieutenant E. H. Thompson, R.E., was responsible for the plotting mechanism design during his period as Research Officer to the Air Survey Committee.

1.1.11
**Barr and Stroud 'Big Bertha' Plotting
Machine Type ZA1.**

In 1928, a request was received from the War Office to design a machine for photographing letters for maps, based on an invention by one of the staff from the Geographical Section. The primary object of the Photonymograph was to produce lettering of the sizes required for affixing to map drawings before their final reduction to the finished scale (Fig. 1.1.12).

A disc carries the master lettering and the letter to be photographed is selected by rotation of the disc. The letter to be copied is illuminated by a lamp and, by means of an optical system with variable magnification, an image of the letter is projected onto photographic paper in the camera.

In 1977, Barr and Stroud Ltd was purchased by Pilkington Brothers Ltd.

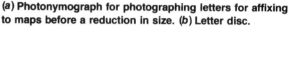
1.1.12
(a) Photonymograph for photographing letters for affixing to maps before a reduction in size. (b) Letter disc.

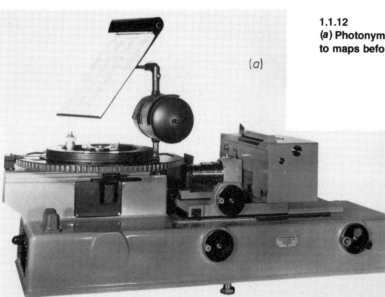

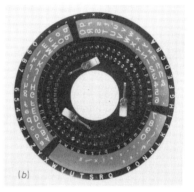

1.2 Chance–Pilkington

During 1957, the optical production activities of Pilkington Brothers Ltd at St Helens, Lancashire, and Messrs Chance Brothers Ltd of Smethwick, Birmingham, were brought together in a new factory at St Asaph, North Wales (Fig. 1.2.1).

A further company, Pilkington P.E. Ltd, was started at St Asaph in 1966 to develop and manufacture optical systems using the most up-to-date machinery and techniques.

Pilkington Brothers Ltd

In 1826, William and Richard Pilkington,

with three local partners, founded the St Helens Crown Glass Company, choosing the site because sand, soda ash, dolomite and limestone were readily available. Coal could be obtained from nearby collieries and the local canal was suitable for transport. When the local partners withdrew their support in 1849, the firm was renamed Pilkington Brothers. At this time, the window-glass industry was almost exclusively concerned with producing crown glass (Fig. 1.2.2).

Henry Chance, writing in 1888, tells us how the crown, or disc, was spun after the initial

1.2.1
Binocular prism mouldings as they emerge from the annealing lehr at the Chance–Pilkington factory.

blowing and shaping stages on the end of an iron rod. He writes:

> A man . . . with a veil before his face, stands in front of a huge circle of flame, into which he thrusts his piece rapidly, meanwhile revolving his ponty [the iron rod]. The action of heat and centrifugal force combined is soon visible. The nose of the piece expands, the parts around cannot resist the tendency . . . the next moment, before the eyes of the spectator, is whirling a thin transparent circular plate of glass which but a few minutes before was lying in the glass pot.

The production of crown glass for windows was superseded during the years 1835 to 1845 by the introduction of sheet glass made by the blown cylinder process. The glass was blown in the form of a cylinder, then split lengthwise and flattened.

Three innovations made it possible to replace the traditional pot furnace by a continuous process. In 1863, the Siemens regenerative furnace was developed to be more economical in the use of fuel. In 1870, the Bievez lehr or cooling oven reduced the annealing time from eight hours to thirty minutes. In 1873, the Siemens tank furnace replaced the

1.2.2
The crown process produced discs of glass about 1·4 to 1·8 m in diameter. This restriction in size, and also the bullion in the centre, dictated the scope for the production of windows. Square panes were cut from a round crown; the numbers give the sizes in inches.

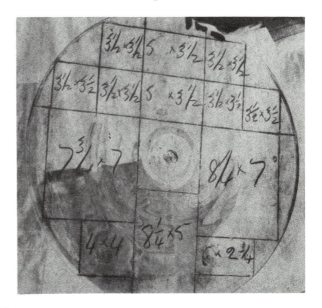

11

traditional pot furnace and made it possible to heat large volumes of glass to the required melting temperature of 1550 °C.

At the same time as the introduction of tank furnaces for the manufacture of sheet glass, a new factory at Cowley Hill, on the outskirts of St Helens, was built for plate glass. By 1878, five years later, as much plate glass was being produced from the new plant as from the original factory, which had been set up in 1773 by the British Cast Plate Glass Company. By 1905, within 30 years of Pilkington's entry into plate glass manufacture, they were the sole British producers, a result of the introduction of numerous innovations and improvements in technology.

Early in the 1920s, Pilkington Brothers cooperated with the Ford Motor Company of America in the development of a flow process, for plate glass manufacture, by which continuous melting in a furnace was combined with the rolling of a ribbon of glass.

In 1923, Pilkingtons introduced the first continuous grinding and polishing machine, which transferred cut plates of glass on a series of tables and moved them through grinders to polishers. In 1935, a twin grinding machine was introduced which could grind the ribbon of glass, about 300 m long, on both sides as it came out of the annealing lehr and before it was cut into plates. This technique for producing high-quality flat glass was a great success, but a 20% glass wastage, with high capital and operating costs, demanded an improved process.

The development of the float glass process was announced in 1959. A continuous ribbon of glass moves out of the furnace and floats along the surface of an enclosed bath of molten tin, in a controlled atmosphere at a high enough temperature, for sufficient time to permit irregularities to melt and allow the two surfaces to become flat and parallel.

The ribbon of glass is then cooled down, whilst advancing across the molten tin, until the surfaces are hard enough for it to be taken out of the bath without marking the bottom surface by the rollers (Fig. 1.2.3).

The advantages of the float glass process are as follows:

(*a*) The liquid surfaces become naturally flat and parallel.

(*b*) The speed at which the ribbon forms is only determined by the melting capacity of the furnace.

(*c*) The manufacturing cost of float glass is only a small fraction of the equivalent polished plate process.

(*d*) The width and thickness of the ribbon can be easily altered within the limits of the

1.2.3
Float glass emerging as a ribbon from the cooling lehr and approaching the automatic warehouse.

float bath and cooling lehr. This reduces warehouse loss by matching the ribbon with the orders to be cut.

(e) The glass produced has a very bright finish and surface damage is much lower than with polished plate.

(f) Because the process is horizontal, the annealing is not a problem.

(g) The finished ribbon, at the end of the lehr, encourages the use of an automatic warehouse for cutting and handling the glass.

(h) The process lends itself to trouble-free runs over a period of at least two years without a stop.

Chance Brothers Ltd

Optical glass cannot be limited to simple compositions. Different purposes require glasses of different refractive and dispersive powers. In 1759, when John Dollond was awarded a patent for his achromatic lens, he had only two glasses, ordinary crown and flint, for his twin components.

A discovery by Pierre Louis Guinard (1748–1824) in 1798, after years of experiments, was that he equalised the density of molten glass by long, continuous stirring which also released air bubbles generated during the process. In 1805, Guinard left Switzerland to join Joseph von Fraunhofer (1787–1826) in Munich, and they produced some glass discs, up to 350 mm diameter, for working into objective lenses.

Guinard's younger son, Henri, was working on the glass production problem with Georges Bontemps and Lerebours in France and, during 1836, also produced a disc of flint glass 350 mm in diameter which was matched with a crown glass in 1843. The objective constructed from these components by Lerebours became the property of the Observatory of Paris during 1849. In 1838, Lucas Chance took out an English patent for Guinard's process following a payment of 3000 francs to Henri Guinard and a similar sum to Bontemps for 'his instructions'.

In 1848, Bontemps left France to join Chance Brothers in England, and a new plant was ordered to make hard crown and dense flint for telescopes, and also soft crown and light flint for photographic cameras. This was the start of the manufacture of optical glass in Birmingham, and London opticians were being supplied early in 1849. By 1851, new dense flints were being made, and an objective disc, 740 mm diameter and weighing about 200 lb (90 kg), was shown at the Great Exhibition and was awarded a Council Medal.

For the next 30 years, a series of astronomical telescope objective discs were made with great difficulty and hazard. Early success meant a large gain, but repetition of failures a heavy loss. Their production, with its slow processes of moulding and annealing, took many months, and if first attempts were not successful the months became years. And all through the long ordeal there was a constant liability to accident. In the later years, large objectives were being replaced by reflectors made from speculum which were much easier to manufacture.

One of the last pairs of discs to be made were 710 mm diameter for Greenwich Observatory, the contract being concluded in 1887. The flint disc was obtained at the first attempt, but it took nearly five years to make the crown disc.

It was not on the production of special objectives, but on the regular satisfaction of opticians for small mouldings, that success depended. Before the Jena glass works in Germany was started, the only makers of optical glass in the world were Chance Brothers and Feil of Paris. There was enough work to keep both firms busy.

In 1862, when Bontemps returned to France, the optical glass department was committed to William Edward Chance, until he retired 10 years later, to be followed by Henry Chance and then by George Chance until 1895. This was a critical time because, over the previous 10 years, a new range of superior optical glasses had been developed in Jena, and the firm of Chance decided that they must attempt manufacture of a similar range at Birmingham or they would lose their business.

Professor Ernst Abbe (1840–1905), Director of the University Observatory, Jena, realised the need to eliminate the secondary spectrum from the ordinary achromatic lens, and that a greater variability was required between the refractive and dispersive powers of lens components. He discussed these requirements, and the means for fulfilling them, in a paper read in 1876.

Achromatic lenses consist of positive lens components made from crown glass, which has a low dispersion, matched with negative components of less power but having a high dispersion. Two colours, such as yellow and blue, can be neutralised, but green is then under-corrected and red and violet are over-corrected.

In order to correct the secondary spectrum, it was necessary to have crown-type glasses with a relatively small dispersive effect at the blue–violet end of the spectrum and flint-type glasses with opposite effects. A careful choice of abnormal glasses will give apochromatic corrections.

In 1880, Dr Otto Schott, a graduate of the school of the Westphalian glass industry, joined Abbe to investigate what optical properties might be imparted to glasses by unusual ingredients. In 1884, Schott, Abbe and Carl Zeiss (1816–88)—the latter had been making microscopes in Jena since 1846—ventured at their own expense to start a glass factory at Jena, and soon they had financial support from the Prussian Government. The first trade catalogue of 1886 listed glasses with 44 different compositions, of which 19 were essentially new. Schott had tabulated systematically the optical properties of all possible compositions through more than a thousand experiments.

After the death of Carl Zeiss in 1888, Professor Abbe became head of the Carl Zeiss works in Jena.

Chance Brothers engaged Arthur H. Lymn as scientific manager in 1897, and he conducted experiments on a large scale, during the next three years, giving particular attention to light flint and dense crown baryta glasses. But he failed to overcome the difficulties. He was replaced in 1900 by Walter Rosenhain, who had a brilliant scientific record, but his research proved to be very costly and failed to solve the problems. In 1909, Samuel Lamb

was appointed manager and was able to make progress with development of the new glasses.

A gas furnace for two 980 mm pots was constructed for continuous working, and the manufacture of large discs up to 700 mm diameter was resumed. Mechanical stirring of the molten glass came into general use at the Chance factory during 1914.

The enormous demand for field glasses and cameras for military purposes stimulated the construction of further furnaces in 1915 and 1916. The difficulties surrounding the production of the new barium glasses were surmounted and British supplies of potash (also carbonate and nitrate of baryta, previously imported from Germany) were rapidly developed.

In conclusion, may be quoted some paragraphs from a paper by Mr F. E. Lamplough published in *Nature*, 27 March 1919:

In considering the position of the industry after the war, it is obvious that there are resources in this country for the manufacture of all the optical glass which will be required by our opticians. Nor need there be any apprehension regarding the ranges of glass which will be available for the use of the lens designer. Without any notable exception, Messrs. Chance Brothers & Co. have been able, by their previous experience and by the work of their research laboratory, to produce glasses which, in their optical constants, cover the full range of glasses mentioned in the Jena list for 1913.

The further development of the optical glass industry would appear to be well provided for in view of the practical research work conducted by the British Scientific Instrument Research Association, recently formed under the direction of the British Optical Instrument Manufacturers' Association. [In 1979, these associations are named the SIRA Institute Ltd and the Scientific Instrument Manufacturers Association (S.I.M.A.).] To maintain the supremacy of the nation in regard to this manufacture, however, it is not only necessary to be able to produce material of good quality, but it is also essential that it should be produced at prices which will compete with those of foreign firms. With the greater time which manufacturers will be able to devote to the subject with this end in view, there should be no difficulty in arriving at a satisfactory solution of this point.

However large the possible output and however perfect the quality of British optical glass, the future of the industry can be assured only if British opticians are able to achieve and maintain supremacy in home and foreign markets by excellence in the design and workmanship of their instruments of precision and by cheapness of manufacture of the more common optical products.

1.3 Dallmeyer

J. H. Dallmeyer Ltd was established in 1860 by John Henry Dallmeyer (1830–83) who was born in Germany and came to England in 1851 to become a naturalised British subject.

In 1860, he patented his first triple achromatic lens which consisted of three cemented doublets. The centre negative doublet was placed in the position usually occupied by the stop, and this had a marked effect in flattening the field. In 1865, he invented the Dallmeyer Rectilinear lens which consisted of two thin pairs of cemented doublets. At the time, this was an excellent all-purpose lens, with a working aperture of $f/8$, and it was used in amateur hand cameras for some 30 years. The Patent Portrait lenses were introduced in 1866.

For some years before the death of his father, Thomas Rudolph Dallmeyer (1859–1906) was responsible for the management of the business and the design of new lenses. The first telephoto lens was designed by T. R. Dallmeyer in 1891, to be followed by the 'Adon' telephoto lens in 1892 which was a very small variable-focus tele-objective. The first edition of the book *Telephotography* by T. R. Dallmeyer was published by William Heinemann in 1899.

The private company J. H. Dallmeyer Ltd was formed in 1892 and, after the death of T. R. Dallmeyer in 1906, Cyril Frederick Lan-Davis continued the pioneer designs of fixed-separation telephoto lenses until his death on active war service in 1915.

The design and improved manufacture of the range of lenses became the responsibility of Lionel Barton Booth who patented the 'Dallon' fixed-separation telephoto lens in 1919. The Dallon telephoto lens was made from two pairs of thin cemented doublets using a special dense barium crown glass developed by Chance Brothers of Birmingham.

The 'Pentac' large-aperture anastigmat lens patented in 1920 consisted of a nearly symmetrical cemented pair, in both front and back, with a single negative element situated close to the iris diaphragm. This unique lens possessed an extreme working aperture of $f/2 \cdot 9$ with good definition and optical corrections.

In order to use this lens design effectively, three sizes of 'Dallmeyer Speed Cameras' were developed, each with a focal-plane shutter. The smallest size, with image of $4 \cdot 5$ cm $\times$ 6 cm, was one of the earliest designs of precision miniature camera.

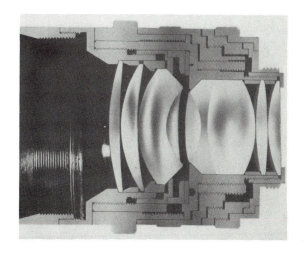

1.3.1
Dallmeyer 'Ultrac' lens for vidicon TV and 16 mm motion-picture cameras. The f/0·98 aperture, and 35° field of view, with 25 mm focal length, gives excellent performance.

15

The early models of enlarging lenses, with f/3·5 and f/4·5 aperture, were reviewed in the *British Journal Almanac* of 1929 and were the first lenses to be designed specially for enlarging.

In 1970, there were nearly 1000 different types of Dallmeyer lens available for industrial, scientific and research applications (Figs 1.3.1 and 1.3.2).

1.4 Dollond and Aitchison

The Dollond and Aitchison Group Ltd was formed in 1927 by the amalgamation of the firms Dollond and Co. and Aitchinson and Co.

Dollond and Co.

John Dollond (1706–61) was the son of Jean d'Hollond who fled to England to escape from persecution, after the Edict of Nantes in 1685, and joined the silk-weaving community in Spitalfields, London. As a boy, John Dollond studied mathematics, optics and astronomy, and by 1750 was known in scientific circles for his wide optical knowledge.

At that time, optical aberrations were being studied by many scientists. Sir Isaac Newton (1643–1727) had stated that refraction was impossible without dispersion and prematurely concluded that an achromatic lens could not be made. In 1729, Chester Moor Hall (1703–71), a barrister by profession and an enthusiastic amateur optician, obtained an achromatic lens by combining together two different glasses of opposite powers, but the significance of the discovery was not fully appreciated or published.

Accounts differ as to the way in which John Dollond learnt of Hall's achromatic lens but,

in the 1764 trial case of Peter Dollond versus the London opticians, it was stated that Dollond heard of the invention in 1750. John Dollond 'made experiments on the different dispersive powers of Water and Glass and from the result of them had attempted to make object-glasses that would represent objects free from colour', but these experiments were unsuccessful.

Dollond tried to find a suitable medium to replace water, and he found that English flint glass possessed a higher dispersive power than crown glass. On the testimony of Jesse Ramsden, an instrument maker of repute, it was in the workshop of George Bass that Dollond found flint glass was the medium he was seeking.

Jesse Ramsden married John Dollond's daughter in 1765 and, as part of a marriage settlement, received a share in the patent for manufacturing achromatic lenses. Ramsden made his first dividing engine in 1766, and this was followed by a superior engine in 1775 which enabled graduations to be divided and cut in metal to every 10 seconds of arc. Details were published in 1777, and this engine was used continuously until his death in 1800 and afterwards by his successors until 1890. In 1775, Ramsden invented a special eyepiece micrometer, for measuring with accuracy the diameters of the exit pupils of telescopes, and was one of the first to apply his positive compound eyepiece to the microscope as a means of reading circle graduations. Among Ramsden's larger instruments were two 36 in diameter circle theodolites made for General William Roy's Trigonometrical Survey of the British Isles from 1791 to 1794. Ramsden was elected a Fellow of the Royal Society in 1786 and in 1795 received the Copley Medal, the senior award of the Royal Society.

In 1757, Dollond discovered that a 25° flint prism and a 29° crown prism gave little resultant deviation but strong dispersion when combined together. He deduced that the focal length of a concave flint to that of a convex crown must be in the ratio 6 : 4, which is the ratio of their dispersive powers. To reduce spherical aberration, he saw the possibility of making the aberrations of any two glasses equal and so, as the refractions of the two glasses were contrary to each other, their aberrations would entirely vanish. In this way, Dollond made his first achromatic lens and tested it by 'evident experiments' (Figs 1.4.1 and 9.1.1).

In 1758, Dollond read and published an important paper entitled 'Account of some Experiments concerning the different Refrangibility of Light' which disproved the conclusion of Sir Isaac Newton.

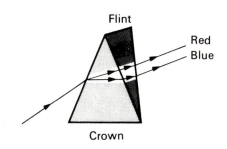

1.4.1
Little deviation and strong dispersion produced by crown and flint pair.

In 1750, Peter Dollond (1733–1820), eldest son of John Dollond, had opened a small optical business, and two years later he was joined by his father at 'The Sign of the Golden Spectacles and Sea Quadrant' in the Strand, London. At this address, John Dollond performed his experiments and made achromatic lenses which were to revolutionise optical instruments. John Dollond was awarded the Copley Medal in 1761 and elected a Fellow of the Royal Society, before dying in the same year.

Peter Dollond had persuaded his father to apply for a patent, which was granted, for the 'new invented method of refracting telescopes, by corresponding mediums of different refracting qualities, whereby the errors arising from the different refrangibility of light, as well as those which are produced by the spherical surfaces of the glasses, are perfectly corrected'. All the early achromatic telescopes had small objectives, owing to the scarcity of good flint glass.

Benjamin Martin, George Adams, John Troughton and other opticians started selling achromatic telescopes in defiance of the patent, and Peter Dollond warned them that if they did not pay him a royalty he would sue them for damages. This provoked 35 London opticians in 1764 to address a petition to the Privy Council requesting the patent be declared invalid, as Chester Moor Hall had invented achromatic lenses, but no action was taken on the petition. In 1766, Peter Dollond sued James Champneys, one of the subscribers to the petition, and the Court agreed that Hall was the inventor, but as he summed up the case Lord Camden said 'it was not the person who locked his invention in his scritoire that ought to profit by a patent for such invention, but he who brought it forth for the benefit of the public'. There were several actions by Peter Dollond for infringement of the patent which remained valid.

Peter Dollond improved his telescope objectives in 1763 by introducing a three-element triplet as a crown, a flint and then a crown glass. The flint was a double concave lens, but it was not easy to obtain a steady flow of good-quality glass and blanks of 100 mm diameter were about the largest size available at the time. Dr William Kitchener, in his *Economy of the Eyes* (1824) describes Peter as 'Father of Practical Optics' (Fig. 1.4.2).

Dollond advertised night glasses, portable equatorials, small reflecting telescopes, compound microscopes, sextants, octants, theodolites and spectacles. Several useful improvements were made to the 'Cuff' microscope, designed by John Cuff (1708–72), which was one of the best of the period (Fig. 1.4.3). A joint was added, which enabled the instrument to be placed in either a horizontal or inclined position, and larger concave and flat mirrors were also fitted.

Another microscope made by Dollond in 1760 was the brass three-pillar Culpeper

1.4.3
Cuff microscope made by Peter Dollond of a general design which continued for a long time from about 1770 and was the prototype for traditional modern designs.

1.4.2
Galilean telescope made by Peter Dollond in about 1765.

microscope, mounted on a rectangular wooden base with drawer, which was designed by Edmund Culpeper (1660–1738) and manufactured well into the nineteenth century (Fig. 1.4.4).

In 1783, Peter Dollond was joined by his brother John, and at this time began to sell telescopes, with brass draw tubes, for the Army and Navy during the Napoleonic Wars.

By 1798, the modifications which had been introduced since Robert Hooke described his microscope in the *Micrographia* of 1665 were all embodied in an instrument known as 'Jones's Most Improved Microscope' (Fig. 1.4.5).

18

1.4.4
Culpeper microscope made by Peter Dollond in 1760.

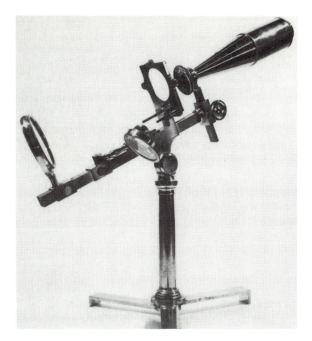

1.4.5
Jones's compound microscope made by Peter Dollond about 1798.

1.4.6
Nautical sundial and sun compass made by George Dollond at St Paul's Churchyard (c. 1850).

In 1820, Peter Dollond died and the business passed to his nephew, George, who collaborated with leading scientists, including Michael Faraday and Sir John Herschel, to produce better lenses. George Dollond died in 1852 and his nephew, another George, took over and was appointed Optician to Queen Victoria. He supplied and fitted all the optical and nautical instruments to the Royal Yacht 'Victoria and Albert'. The second George Dollond died in 1866 but the business of Dollond and Co. continued to grow (Fig. 1.4.6).

Aitchison and Co.

James Aitchison (1860–1911) was apprenticed to a firm of opticians in High Holborn and, at that time, sight testing was carried out by rule-of-thumb methods. He developed scientific methods for sight testing and in 1889 set up business in Fleet Street. He pressed for proper qualifications and training for ophthalmic opticians. He passed the first examination held by the Worshipful Company of Spectacle Makers (Fig. 1.4.7).

When James Aitchison died in 1911, his son Irvine G. Aitchison (1888–1965), Barrister at Law, who had an interest in the Wray Optical Company, assumed responsibility for the firm Aitchison and Co. which was primarily engaged in the provision of spectacles through branch shops.

In 1927, the firm was amalgamated with Dollond and Co. to form Dollond and Aitchison Ltd which Irvine Aitchison controlled until 1964. At the headquarters of Dollond and Aitchison Group Ltd, Yardley, Birmingham, there is a small museum in which there are many examples of historical products made by the company (Fig. 1.4.8).

1.4.7
Field glass, consisting basically of two Galilean telescopes, made by Aitchison and Co. in 1900.

1.4.8
(a) Museum at Dollond and Aitchison Group Ltd, Yardley, Birmingham. Telescopes and microscopes made by Dolland, and also spectacles made by Aitchison, are on show. **(b)** Dollond telescopes and a wheel for measurement of a mile.

Joseph Jackson Lister (1786–1869), father of Lord Lister, during 1829 read a paper to the Royal Society which described improvements to compound microscope lenses, and his introduction of a hemispherical front lens, and these have formed the basis for modern achromatic microscope objectives (section 3.1).

J. J. Lister had two nephews, Richard and Joseph Beck, who were to learn optical manufacturing techniques and be founders of the company R. and J. Beck Ltd which traded continuously for more than 100 years.

Richard Beck was apprenticed to James Smith, an instrument maker. By 1842, Richard Beck was producing the world's first microscope with compound achromatic objectives and was the inventor of a thin glass vertical illuminator which made possible the examination of polished metal surfaces under high powers for metallurgy. In 1847, he entered into partnership with Smith to trade as Smith and Beck.

Joseph Beck served his apprenticeship with Troughton and Simms, and then joined the firm Smith and Beck in 1851. He became a partner in 1857 and in 1867, upon the retirement of Smith, the name of the firm was changed to R. and J. Beck (section 1.9).

As the years passed, design and production became the province of Joseph, whilst Richard developed applications for microscopy and was the author of several books.

Before the middle years of the nineteenth century, when intense industrial rivalry developed, scientific research was carried out by gifted amateurs and men of property who could devote time, skill and resources to the pursuit of knowledge. Discoveries were shared at informal meetings in societies and discussion groups where scientific progress was the main driving force. The enthusiasm of these pioneers was matched by dedicated craftsmen exploiting unique skills to produce custom-built instruments and equipment. Microscopists would meet Beck, to discuss their preparations and for admiration and comment, under hospitable conditions extending well after normal business hours.

Since the 1920s, there has been close co-operation with the Scientific Instrument Research Association (SIRA Institute Ltd) leading up to the Eros M.T.F. systems now widely used throughout the whole optical industry.

During 1926, Beck microscope objectives, mechanical slits and a lamp condenser were supplied to De Forest Phonofilm Ltd for mounting into 35 mm motion-picture projector soundheads ready for the first sound-on-film pictures which were then being made in Hollywood.

At that time, Conrad Joseph Beck was responsible for instrument design, supported by Mr Wybourne on optical design, and David W. Long a mechanical designer who became the optical designer in 1939. During this period, Beck optics were supplied to motion-picture manufacturers for reproduction and also recording of sound-tracks (section 4.3).

British Acoustic Films Ltd, a subsidiary of the Gaumont–British Picture Corporation, placed substantial orders for optics to be fitted on Gaumont projector soundheads. At the same time, sound-recording cameras were developed using a variable-area sound-track. The recording depended on an image moving across the film by means of a very small mirror, oscillating at up to about 10 kHz, immersed in oil which magnified the movement and provided dampening of the vibration. A new design of Gaumont–Kalee sound-recording camera with Beck optics was produced during 1948 which increased the available light on the film (Fig. 1.7.5).

Further development was required for push–pull sound tracks which were introduced to assist in reducing distortions and losses which occur during mixing, dubbing or re-recording (Fig. 4.3.1).

In 1968, R. and J. Beck Ltd was acquired by The Ealing Corporation, South Natick, Massachusetts, U.S.A. and became Ealing Beck Ltd with a factory at Watford, England. Many of the components and instruments in the optics section of the Ealing catalogue are manufactured at the Beck factory.

1.6 Grubb–Parsons

This is a story of two fathers, Thomas Grubb, F.R.S. (1800–78), and William Parsons, F.R.S., Third Earl of Rosse (1800–67), and their sons, Sir Howard Grubb, F.R.S. (1844–1931), and Sir Charles Parsons, O.M., K.C.B., F.R.S. (1854–1931).

Thomas Grubb became interested in optics and constructed a small observatory at Charlemont Bridge, in which he installed a 9 in reflecting telescope, and near which he set up his workshop. In 1834, he made a 15 in reflecting telescope for the Armagh Observatory, and later a 12 in refractor for Dunsink Observatory which is still in regular use. At the Dublin Exhibition of 1853, Thomas Grubb displayed a 12 in refractor which attracted attention, and as a result he was entrusted with the design and manufacture of the 'Great Melbourne Telescope'. This was a 48 in reflector, with a speculum mirror, which was completed in 1868. In 1954, this instrument was moved to the Commonwealth Observatory at Mount Stromlo, Australia, where it was rebuilt and fitted with a new 50 in (1270 mm) low-expansion glass mirror (Fig. 1.6.1).

William Parsons, the eldest son of the second Earl of Rosse, entered Parliament in 1823 but resigned his seat in 1834 so that he could devote his time to philosophical pursuits. The professional telescope makers of the period kept their methods of casting, grinding and polishing of speculum a very close secret, but in 1828 William Parsons published his first paper giving results of his experiments. He concentrated his efforts on reflecting telescopes and in 1830 described the casting of five specula of 15 in diameter with ribbed backs. A paper to the Royal Society in 1840 described his methods for making speculum mirrors up to 36 in diameter.

At the age of 41 years, William Parsons succeeded to his father's title and became the third Earl of Rosse. In 1842, a 5 ft (1·83 m) diameter speculum mirror was cast for his telescope which, with a 54 ft (16·46 m) tube, was completed in 1848 (Fig. 1.6.2).

With this telescope, thousands of observations were made, including micrometre

1.6.1

The Great Melbourne Telescope made by Thomas Grubb and his son Howard. This was a 48 in reflector and for many years was the largest equatorially mounted instrument in the world.

measurements, by Lord Rosse and visiting astronomers. Some of the earliest work on line spectra of the nebulae was reported by him.

Sir Howard Grubb chose an optical career and at the age of 20 years decided to join his father to help with building the Great Melbourne Telescope. By the time his father died in 1878, Howard was a competent maker of large telescopes.

At the International Astronomical Congress in Paris during 1888, it was decided to organise a systematic survey of the heavens in both hemispheres and to prepare an International Astrographic Catalogue. Ten identical telescopes, each consisting of a 13 in photographic and 10 in visual telescope together on an equatorial mounting, were ordered and seven were made by Sir Howard Grubb's firm in

1.6.2
Lord Rosse's Telescope at Birr Castle, County Offaly, Ireland.

Dublin. These instruments were set up in Greenwich, Oxford, Cork and Mexico in the northern hemisphere and also at Cape Town, Sydney and Melbourne south of the equator. Some are still in use today.

Many astronomical telescopes were made, as well as a range of surveying instruments and periscopes for early British submarines. The factory was moved from Dublin to St Albans, Hertfordshire, in 1915.

Sir Charles Parsons, youngest son of the third Earl of Rosse, is well known for his work on steam turbines but, in 1889, he became interested in making searchlight mirrors and developed methods of bending glass blanks up to 2 m diameter. In 1920, he acquired all the shares of the Derby Crown Glass Company and an early product was a 27 in (686 mm) flint glass disc for Sir Howard Grubb who was engaged on the Johannesburg refractor.

In 1925, at the age of 81 years, Sir Howard Grubb decided to retire and, by a happy arrangement, Sir Charles Parsons was able to take over the Company and transfer it to Walkergate, Newcastle upon Tyne, near to his turbine works.

The new company was named Sir Howard Grubb, Parsons and Company Ltd, and the first large telescope made at the Newcastle works in 1928 was the 36 in (914 mm) reflector for the Royal Observatory, Edinburgh (Fig. 1.6.3).

The Royal Observatory, established in the 1830s, combines under one roof an establishment of the Science Research Council and a University Department which is closely integrated with the Astrophysics and Seismology groups in the Observatory.

During the 1960s, when George Sisson O.B.E. was Managing Director, there was a close association and interchange of technical information with Hilger and Watts Ltd.

23

1.6.3
The 36 in reflector with attached spectroscope at the Royal Observatory, Edinburgh.

1.7 Rank Precision Industries and Rank Xerox

Roger Bacon (*c.* 1220–92) is said to have invented the magic lantern. Details were published by Giacomo Della Porta (*c.* 1537–1602) in his book *Magia Naturalis* indicating difficulties due to lack of suitable light sources, and a demonstration during 1665 was described in Samuel Pepys's *Diary* (1668). A knowledge of photography and the phenomenon of persistence of vision was needed in order to develop the motion-picture projector.

The first regular public exhibition of moving pictures lasting 20 seconds was during the spring of 1896. Because at least 16 photographs per second are needed for motion pictures, a short exposure time on emulsion-coated film is essential, and this was not possible in the early days of photography. The first motion-picture camera, made in 1888 by Professor E. J. Marey, used a roll of sensitive paper which was arrested in its movement by an electromagnet. As the paper rolls were only 4 m in length, there were not more than 40 exposures on each roll. The prototype motion-picture cameras made in the years 1888–91 were similar to the successful cameras of 1896 which used celluloid perforated film, moved frame by frame, with a shutter masking the film movement.

The cinematograph was one of the last inventions of the Victorian age and the 'picture

24

palace' satisfied one of the social needs of the time. Many people deserve credit for this invention, but an Englishman, William Friese-Greene (1855–1921), applied for a patent in June 1889 describing a motion-picture camera which was 'the prior patent of the world' in this particular field. The patent specified the use of photographic sensitive film and embodied every essential feature with the exception of sprocket wheels to give intermittent motion.

The patent was upheld in the American courts against a counter-claim by Thomas Alva Edison whose patent of 1891 for his 'Kinetoscope' viewer was only granted in the United States as he failed to apply for a patent in England.

The Edison 'Kinetoscope' used a continuous film movement and momentary light flashes from the slit in a revolving shutter which was placed between the film and an electric lamp. The film was 35 mm wide and four perforations in the margin of each picture were used for registration. This film size has been adopted as an international standard.

The Kinetoscope was a one-person viewing cabinet and the continuously moving, momentarily illuminated principle was unsuitable for projection.

In 1896, Robert W. Paul, an instrument maker in London, was asked to make copies of the Edison Kinetoscope, and several hundred machines were made. By an improvement to the shutter, and an intermittent film motion using two seven-star Maltese crosses, a projector was designed to throw pictures onto a screen and this was named the 'Theatregraph'. The first film shown with the Theatregraph in February 1896 was followed by a demonstration, before the Royal Institution, of the intermittent film motion which was used on subsequent motion-picture projectors.

In the same month, Monsieur Treway demonstrated in London 'Le Cinématographe' which had been invented by the brothers Louis and Auguste Lumière of Lyons. The first film trade show was held in 1898 and A. C. Bromhead, British representative of M. Leon Gaumont, was the projection-machine operator.

A Frenchman, Eugene Lauste, working in a studio at Brixton during 1900, was the first to reproduce sound on film. Others such as Edison and Graham Bell, succeeded in recording sound on photographic plate but could not reproduce it. Edison had to wait for George Eastman (1854–1932) to invent the long-length roll of celluloid film before he could operate his sound machine.

In 1904, A. C. Bromhead, who later became Chairman of the Gaumont–British Picture Corporation when it was formed in 1927, was the first to operate a permanent film show, 'The Daily Bioscope', in a Bishopsgate room with space for about a hundred people. During the next 10 years, films increased in length from a few hundred feet to the full-length fiction film on which the world of cinema entertainment was to prosper. By 1914, silent films made in Hollywood, California, were flooding the motion-picture market and millions of people were being entertained at the cinema. It was not until the late 1920s that British films had acquired character and were made with competence.

During 1926, various experiments with sound on film were being made in England by British Acoustic Films Ltd and also by De Forest Phonofilm Ltd. They were ready for the first full-length semi-sound production from Warner Brothers in 1927, which had sound tracks on gramophone records. The Western Electric Corporation and the Radio Corporation of America led the way in supplying sound reproducing equipment to American cinemas.

Rank Precision Industries Ltd was formed in 1948 to group together the major British manufacturers and distributors of film studio and theatre equipment. The most important manufacturing units were:

(a) Kershaw factory (previously A. Kershaw and Sons Ltd) at Leeds (motion-picture projectors, lenses and arc lamps).

(b) British Acoustic Films Ltd (later renamed Rank Precision Industries (B.A.F.) Ltd) with factories at Woodger Road, Shepherds Bush, London, and at Mitcheldean, Gloucestershire (sound-recording and reproducing equipment).

(c) Taylor, Taylor and Hobson Ltd with factories at Leicester (camera lenses and engineering instruments).

At that time, 8 mm and 16 mm motion-picture cameras and projectors were made under licence from the Bell and Howell Company, Chicago, but 35 mm sound-recording and reproducing equipment and also projectors and arc lamps, for studios and cinemas, were designed and manufactured within the Rank Precision Industries group of factories.

J. Arthur Rank (1888–1972), President of The Rank Organisation Ltd, who became The Lord Rank, L.L.D., J.P., in 1957, was a multi-millionaire from his family flour-milling business which was founded in the nineteenth century. As a devout Methodist, he thought that religious films, as simple well made productions, would be an effective way of teaching the Gospel in churches. It is interesting to note that the first contact Mr Rank made with the film industry in the early 1930s was through his Religious Film Society who purchased a substantial quantity of Gebescope 16 mm sound projectors manufactured by British Acoustic Films Ltd (Fig. 1.7.1).

In order to make religious films especially for Sunday Schools, Mr Rank formed G.H.W. Productions Ltd in 1935 and purchased Gate Studios at Elstree. For economic reasons, these film productions were closed down 10 years later, and Gate Studios was purchased from Mr Rank by Rank Precision Industries Ltd in 1953 for the manufacture of Cinemascope screens and frames. Part of the land was sold to Rank Xerox Ltd, in 1957, as a site for their first factory, which was for producing Xerographic toning powder and for vacuum coating selenium onto copying machine drums and plates.

A short amateur documentary film entitled 'The Turn of the Tide' had been made in 1934 about the social conditions in England at that time. Demonstrations against a reduction in unemployment pay for the $2\frac{1}{2}$ million people out of work were compared with the affluence displayed at Henley, Lords, Ascot and Cowes. Mr Rank wanted this film released on the

1.7.1
Joseph Arthur Rank (1888–1972), Founder and President of the Rank Organisation Ltd, was made First Baron Rank in 1957 (photograph 1952).

Gaumont–British Cinema circuit and, because of difficulties he encountered, he determined to gain control over his own film productions including the way in which films were distributed and shown in cinemas. The developments which led to the formation of the Rank Organisation were never, as far as Mr Rank is concerned, for financial gain, as he already had a large private fortune and film production had always offered the probability of financial loss.

Mr Rank wanted stability and fair dealing so, as a step towards removing inefficiency and improving the quality of films, he built Pine-wood Studios in 1935 and acquired an interest in the British part of Universal Pictures which was merged with his company General Film Distributors Ltd.

In 1941, the Gaumont–British Picture Corporation, with 60 companies and a six-figure overdraft, was in a weak financial position and the 20th Century–Fox Film Corporation, an American film production company with a substantial shareholding in Gaumont–British, wanted to acquire control. Through a complicated series of negotiations, to prevent American domination of the industry, Mr Rank purchased the controlling shares of the Gaumont–British Picture Corporation. The framework of the Rank Organisation under British control had been formed, including ownership of the subsidiary companies, British Acoustic Films Ltd and G. B. Kalee Ltd. As part of the agreement, Mr Rank invited Mr Spyros K. Skouras, President of the 20th Century–Fox Film Corporation, to join the Board of Gaumont–British and Mark Ostrer became a Joint Managing Director.

A month after Rank became Chairman of Gaumont–British, Oscar Deutsch, the founder of Odeon Theatres, died and Deutsch's trustees offered the Odeon interests to Rank, who agreed to buy them and so became Chairman. John Davis, who had joined Odeon in 1938 at the age of 32, became a Joint Managing Director in 1941, Managing Director in 1948, and Chairman of the Rank Organisation in 1962.

Thomas A. Law (1906–67), an electrical and radio engineer with extensive knowledge of the sound-on-film recording and reproduction industry, joined British Acoustic Films Ltd in 1929 as Chief Engineer and in 1945 became a Director of Taylor, Taylor and Hobson Ltd, G. B. Kalee Ltd and British Acoustic Films Ltd. He was appointed Production Director of Rank Precision Industries Ltd in 1948, when the Company was formed, and became Managing Director in 1951. The first 10 years under his management, from 1948 to 1958, were stable and prosperous (Fig. 1.7.2).

The large-volume production of projectors, lenses, sound equipment, screens and frames, consequential to the introduction of 'Cinema-scope' anamorphic films by the 20th Century–Fox Film Corporation from 1952 to 1956, and the negotiations leading up to the formation of Rank Xerox Ltd in 1956 were major achievements during this period.

Tom Law was the first Managing Director of Rank Xerox Ltd and, because of the increasing importance of Xerographic products, he resigned from Rank Precision Industries Ltd in 1963 and was appointed to the Board of the Rank Organisation Ltd in 1964.

1.7.2
Thomas A. Law (1906–67) was the first Managing Director of Rank Precision Industries Ltd from 1951 to 1964 and first Managing Director of Rank Xerox Ltd from 1956 to 1967 (photograph 1952).

Kershaw factory at Leeds

In 1888, Abram Kershaw (1861–1929), a young Halifax instrument maker, started business on his own account and, in 1898, Cecil Kershaw started work with his father making mirror galvanometers and camera brass-work.

In 1905, the Soho Reflex Camera was designed and manufactured by A. Kershaw and Sons Ltd and this was followed in 1910 by the production of 35 mm cinema projectors to compete with imported American machines. Because the distributors did not wish the manufacturers name to appear on the projector, it was labelled Kalee, signifying *K*ershaw–*A*bram–*LEE*ds, and so a famous trademark (later changed to G.B.–Kalee) was founded.

Many thousands of Kalee projectors from Model 1 to Model 12 were made from 1910 to 1947 when a new machine was introduced, known as the GK21, shortly followed by GK20, 19 and 18 of the same family. This new range of projectors, with large-aperture lenses, was very suitable for 'Cinemascope' anamorphic film projection and large numbers were made from 1952 to 1956 (Fig. 1.7.3).

Binoculars, telescopes, periscopes and cameras, of their own design, had been in continuous production since the Company was founded in 1888.

Rank Precision Industries (B.A.F.) Ltd, Shepherds Bush, London W12, and Mitcheldean, Gloucestershire

Originally named British Acoustic Films Ltd, the company was formed in 1925, and amalgamated with British Talking Pictures Ltd to manufacture sound-reproducing equipment for fitting to 35 mm projectors which had been originally designed for silent films. The first-ever full-length talking picture 'The Jazz Singer', was produced by Warner Brothers in 1928 featuring Al Jolson.

Although Gebescope 16 mm sound projectors had been designed and manufactured by British Acoustic Films Ltd in the 1930s, it was an agreement made between Mr J. Arthur Rank and the Bell and Howell Company, Chicago, in 1945 that was to provide the large sales turnover and a substantial part of output for the next 15 years. Because British Acoustic Films Ltd was a subsidiary of the Gaumont–British Picture Corporation, the equipment manufactured under this agreement was marketed with the name GB–Bell and Howell.

The author (D. F. Horne), who had been employed as Group Production Engineer when Rank Precision Industries Ltd was formed in 1948, was appointed a Director of British Acoustic Films Ltd in 1953 and joined J. Arthur Rank, H. R. A. de Jonge, T. A. Law and Mark Ostrer on the Board (Fig. 1.7.4).

Manufacture and assembly of 8 mm and 16 mm GB–Bell and Howell cameras and projectors was centred at Mitcheldean under

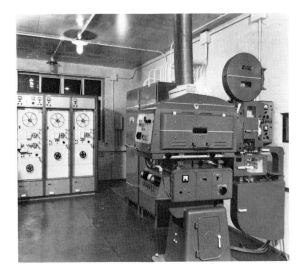

1.7.3
GK20 projector and optical soundhead with dubbing equipment at Pinewood Studios, Iver Heath. The projection equipment was made at the Kershaw factory, Leeds and the optical sound-reproducing equipment was made by British Acoustic Films Ltd at Woodger Road, Shepherd's Bush, London.

28

the management of F. Wickstead, whilst design and production of sound-recording and reproducing equipment was based at Shepherds Bush. S. Pratt was Chief Engineer responsible for research and development until 1960, when he moved to Mitcheldean, and by 1973 had become Director of the Engineering Divison of Rank Xerox Ltd.

In 1952, a special 16 mm fast-pull-down telerecording camera was ordered by the British Broadcasting Corporation to be suitable for the simultaneous recording of the television picture and sound in correct relative positions on 16 mm film. This was necessary at

1.7.4
Photograph taken at Bell and Howell Factory, Chicago, 1949.
Standing (*left to right*): B. E. Stechbart (Chief Engineer, B&H), J. Lawrence (Public Relations Director, J. Arthur Rank Organisation, New York), E. Schimmel (Vice President, International Division, B&H), R. Nerhus (Resident Engineer, B&H, England), D. F. Horne (Group Production Engineer, Rank Precision Industries Ltd), C. E. Phillimore (Vice President and Works Manager, B&H), P. Wagner (Director of Public Relations, B&H), E. Lamb (Manager, International Division B&H), R. S. Benjamin (J. Arthur Rank Co., New York), M. G. Townsley (Assistant Vice President in Charge of Design Engineering, B&H). Seated (*left to right*): A. S. Howell (Chairman, B&H), H. McDermott (Director, B&H), J. Arthur Rank, C. H. Percy (President, B&H).

that time for recording outside broadcast sporting events, such as cricket, when the machine frequently had to start and stop. The television picture-and-sound broadcast of the Coronation of Her Majesty Queen Elizabeth II on 2 June 1953 was recorded on this camera, and a copy of the seven-hour historic film, after processing, was flown to America for rebroadcasting on the same day. This was one of the earliest British telerecordings (see section 4.3).

Before the development of video recordings on magnetic tape, the only way to record a television picture was to take a photograph of each frame, at 25 frames per second, and move the film during the very short time between the end of one television picture scan and the start of the frame for the next picture. Very high accelerations were demanded from the claw and film sprocket holes (Fig. 1.7.5).

Because a large part of the Rank Precision Industries Group turnover was from the sale of equipment supplied to the film industry, and this was declining due to growth of television, it was essential that a range of products with new markets was introduced.

Through a series of lucky circumstances during 1954, it was found that the Haloid Company in the United States was seeking a European partner for manufacturing their xerographic equipment. In May 1955, there was a demonstration in London of a hand-operated xerographic flat-plate apparatus, and in 1956 a new British company to be known as Rank Xerox Ltd was formed to market xerographic equipment, in which the American company had a 50% interest. T. A. Law was appointed Managing Director of Rank Xerox Ltd and the author (D. F. Horne) was invited to join their Management Board to be responsible for production of the document-copying equipment at the Woodger Road factory, Shepherds Bush.

1.7.5
(a) Special fast-pull-down camera mechanism and sound-recording optics for 16 mm film telerecording made by British Acoustic Films Ltd, Woodger Road factory, Shepherd's Bush, London, in 1952. Picture and sound were photographically recorded simultaneously from a special television monitor. The equipment was used by the British Broadcasting Corporation at Alexandra Palace for recording outside broadcasts and sporting events. (b) 16 mm fast-pull-down telerecording camera fitted to television receiver equipment made by Cinema–Television Ltd.

(a)

1.7.6
(a) Rank Precision Industries Ltd factory at Woodger Road, Shepherd's Bush, London, where copying machines were first made for Rank Xerox Ltd in 1956. Gaumont–Kalee sound-recording and reproducing equipment was manufactured in this factory.
(b) Assembly shop at Woodger Road, Shepherd's Bush, in 1961, for Rank Xerox 'Copyflo' machines. This is where Rank Xerox manufacture started.

(b)

31

The 1956 Annual Report of Rank Precision Industries Ltd states:

Arrangements have been made for the manufacture by Rank Precision Industries Ltd in the United Kingdom of the apparatus for Xerography and there is every indication that a very considerable world market exists for these products.

Manufacture of xerographic document-copying equipment was substantially carried out at the Woodger Road factory, Shepherds Bush, until 1961, when production was transferred to Mitcheldean where adequate space was available for expansion (Fig. 1.7.6).

Nobody could have foreseen that in the 20 years from 1956 to 1975 the turnover of Rank Xerox Ltd would increase from nil to £612 million, with a profit before taxation of £145 million a year. This outstanding performance would have been impossible without the vision and drive of Joseph C. Wilson, President of the Xerox Corporation, U.S.A., and Sir John Davis, Chairman and Chief Executive of The Rank Organisation, who were ready to take risks with a very large capital investment on a small range of products in an entirely new market.

In 1970, Fred Wickstead, General Manager of Mitcheldean factory, was appointed to the Board of Rank Xerox Ltd as Director of Production and Supplies and then, in 1971, became Vice President of Manufacturing and Logistics with the Xerox Corporation in America.

Taylor, Taylor and Hobson Ltd

In 1886, the young brothers Thomas Smithies Taylor and William T. Taylor (1864–1957) started the Taylor–Hobson Company and soon achieved an international reputation for advanced design and technical quality. In 1893, the manufacturing rights for the 'Cooke' photographic lens were purchased by them from T. Cooke and Sons Ltd of York where the lens had been invented by H. Dennis Taylor (1862–1943) who was Manager of the Cooke optical workshops.

With the increasing demand for special-purpose lenses, it became necessary to engrave

their mounts and, as no suitable engraving machine was available, William Taylor decided to design and build a machine for this purpose.

In 1914, Arthur Warmisham, a mathematician who had been a student at Manchester University under Rutherford (1871–1937) joined the Company and was responsible for optical research and development until he retired in 1950. In the early 1920s, H. W. Lee, a member of Warmisham's team, designed the famous *f*/2 Speed Panchro lens for 35 mm motion picture cameras which became the most popular large-aperture photographic lens used in the Hollywood film studios.

When Technicolor in America first designed their three-colour system for motion pictures, in which one lens was required to record three separate images of different colours on three films simultaneously, it was the Taylor–Hobson design team who perfected the formation of three images of exactly the same size and supplied lenses for this colour process.

From 1950 until 1980, G. H. Cook was responsible for optical research and development. During this period, the Varotal range of television and studio zoom camera lenses was designed and manufactured in large quantities.

In 1931, William Taylor sold his shares in the Company to the Bell and Howell Company, Chicago, who were the sole distributors of Taylor–Hobson lenses in America. This enabled Bell and Howell to have access to the Taylor–Hobson experience of lens design and production which they needed at their own factory in Chicago to complete their facilities for manufacturing 35 mm, 16 mm and 8 mm motion-picture cameras.

Mr J. Arthur Rank purchased the shares in 1946 as part of an agreement between British Acoustic Films Ltd and the Bell and Howell Company, in which Bell and Howell products would be manufactured under licence in England.

Precision measuring instruments such as the 'Talysurf' for surface finish, 'Talyrond' for roundness measurement, workshop microscopes and alignment telescopes have all formed part of the design programme in which an instrument was developed for the manufacture

1.7.7
Rank Precision Industries group visit to the Production Engineering Research Association, Melton Mowbray, in 1954. (See the attached key and accompanying table for details of the people shown and positions held.)

No.	Name	Factory	Position in Company
1	F. Willbond	T., T. & H., Leicester	Mechanical Devt
2	A. V. Wood	P.E.R.A.	—
3	M. H. Taylor	T., T. & H., Leicester	Managing Director
4	F. W. Wickstead	Mitcheldean	Works Manager
5	A. Taylor	T., T. & H., Leicester	Chief Planner
6	F. W. Court	Mitcheldean	Chief Inspector
7	G. Burnham	Woodger Road	Chief Draughtsman
8	A. S. Pratt	Woodger Road	Chief Engineer
9	Miss Bracken	T., T. & H., Leicester	Secretary to Managing Director
10	V. R. York	A.K.S., Leeds	Chief Planning Engineer
11	Mr Tinsley	P.E.R.A.	—
12	C. Wright	T., T. & H., Leicester	Tool Design
13	R. Baker	Mitcheldean	Works Superintendent
14	J. E. Hambrey	Woodger Road	Works Manager
15	J. A. Stafford	T., T. & H., Leicester	Director & General Manager
16	J. Kellet	A.K.S., Leeds	Planning Dept
17	A. Hagon	T., T. & H., Leicester	Production Manager
18	T. W. Clifford	T., T. & H., Leicester	Director
19	J. A. Mills	T., T. & H., Leicester	Instrument Devt
20	D. F. Newstead	A.K.S., Leeds	General Manager
21	Dr Galloway	P.E.R.A.	Director of Research
22	H. Lustig	Woodger Road	Production Controller
23	W. Blaitch	Mitcheldean	Chief Tool Engineer
24	G. Rawstron	T., T. & H., Leicester	Machine Tool Devt
25	H. Stapleton	Woodger Road	Assistant Chief Draughtsman
26	J. Knox	Mitcheldean	Production Engineer
27	A. Munslow	Woodger Road	Production Engineer
28	D. F. Horne	Woodger Road	Director, Rank Precision Industries (B.A.F.) Ltd
29	Dr Pelton	A.K.S., Leeds	Technical Manager, Opt. Div.
30	N. Addison	A.K.S., Leeds	Chief Designer
31	A. D. Snutch	T., T. & H., Leicester	Production Engineer
32	R. C. Spragg	T., T. & H., Leicester	Electronic Devt

of lenses and was then produced for sale.

To a very large extent, the success with these measuring instruments was due to the initiative of R. E. Reason, O.B.E., F.R.S., D.Sc., during the years from 1944 to 1970 when he was responsible for research in the Taylor–Hobson metrology laboratory.

From 1948 to 1956, Mark H. Taylor, son of Thomas Smithies Taylor, was a Director of Rank Precision Industries Ltd and started the Central Research Laboratories at Shepherds Bush which led to negotiations prior to the formation of Rank Xerox Ltd (Fig. 1.7.7).

In July 1968, the Taylor–Hobson Division of Rank Precision Industries Ltd merged with Hilger and Watts Ltd which had been formed by the amalgamation of Adam Hilger Ltd with E. R. Watts and Son Ltd in 1948 and joined by W. Ottway and Co. Ltd in 1961 and by Wray (Optical Works) Ltd in 1962.

Adam Hilger Ltd

Adam Hilger (1839–97) and his brother Otto (1850–1902) founded Adam Hilger Ltd, in London during 1874, after both of them had served apprenticeships under their father who was Master of the Mint at Darmstadt.

In the early years of the Company, orders were received from Lieutenant-Colonel Archibald Campbell, later to become Lord Blythswood, who had a life-long interest in the physical sciences and was elected a Fellow of the Royal Society in 1907.

In 1888, Otto Hilger worked directly for Lord Blythswood in the construction of a dividing machine, for the ruling of diffraction gratings, but Lord Blythswood died before the work was finished. The unfinished machine was presented to the National Physical Laboratory in 1908 where it was completed and used for ruling gratings during the next 30 years. The Blythswood ruling machine is now on display in the Science Museum at South Kensington (see section 1.8 of *Dividing, ruling and mask-making*).

When Adam Hilger died in 1897, Otto left his work with Lord Blythswood to take charge of the firm, and in 1898 employed a young assistant, aged 22 years, named Frank Twyman who was to exert a far-reaching influence on the company and its future development. At this time, the main output of Adam Hilger Ltd consisted of spectroscopes and optical components for other instrument companies.

Frank Twyman (1876–1959) decided that an expansion of the range of instruments was essential and in 1902 moved the production to much larger premises in Camden Road, London, where the Company was based until 1970. Otto Hilger died in 1902 and in 1904 Frank Twyman was appointed Manager of the business.

In 1904, Twyman began the development of instruments for spectrochemical analysis and their application to industry. At that time, although Ångström in 1855, and Kirchhoff and Bunsen in 1859, had demonstrated spectra characteristics, and Lockyer had patented a qualitative method of analysis with the spectroscope in 1873, the technique had not been used by chemists to aid their analytical work.

Twyman designed the first wavelength spectrometer, which had a helical drum from which it was possible to read the selected wavelength of light. In 1909, the first fixed-adjustment quartz spectrograph was sold to The American Brass Company who published a paper on their use of the spectrograph for production control. In 1913, Twyman produced the first infrared spectrometer, and in 1919 the first commercially manufactured vacuum spectrograph for use by the distinguished physicist, Niels Bohr. The vacuum ultraviolet is a useful region of the spectrum because spectral lines are less crowded and in this region the lighter elements, such as gases and gaseous isotopes, have their strongest lines.

No history of Adam Hilger Ltd is complete without reference to absorption spectroscopy and the famous 'Spekker' absorption meter which was first produced in 1937. This instrument was sold in large numbers and, after modification, included a facility for fluorimetric determinations.

Frank Twyman was elected a Fellow of the Royal Society in 1924 and was Managing Director of Adam Hilger Ltd until he retired in 1946, shortly before the formation of Hilger and Watts Ltd in 1948. He was the author of *Prism and Lens Making* published in 1943, which became a standard textbook on the production of optical surfaces.

The increased use of spectroanalytical instruments for production control demanded an expansion in research, development and production over the 20 years from 1948 to 1968. Spectrographs, employing photographic detection systems, were unable to match the production speeds required by customers, so in the early 1950s a direct-reading grating spectrometer was developed, and the first production model was installed during 1954. This was the first direct reader to employ a digital readout system and soon afterwards a printer was included. A year later, a direct-reading attachment was fitted to the medium quartz spectrograph.

In 1957, Hilger and Watts Ltd produced their first direct-reading vacuum spectrometer, specifically designed for the analysis of steel, including the determination of elements such as carbon, sulphur and phosphorus, which were outside the scope of air-path instruments. Two years later, the 'Polyvac' series of direct-reading spectrometers was available for sale to industrial laboratories. The most comprehensive equipment will measure concentrations of up to 60 elements and provide results in less than two minutes.

E. R. Watts and Son Ltd

Edwin R. Watts founded the company in 1856, after serving an apprenticeship with an instrument maker, and received his first order from Negretti and Zambra for a mining dial. In 1873, the business moved to larger premises in Camberwell, South London, where it expanded until production was transferred to a new factory at Debden in Essex over the period from 1955 to 1970.

In 1886, Frank C. Watts entered his father's firm and George W. Watts followed in 1888. At this time, the company was designing and making heliographs, theodolites and levels, along with other surveying equipment. After 45 years service, Edwin Watts died and was succeeded by Frank Watts in 1901.

In 1904, George Watts completed their first

dividing machine for ruling circular scales (Fig. 1.7.8) and this was in regular use until 1970 when it was presented to the Science Museum, South Kensington (see section 1.5 of *Dividing, ruling and mask-making*).

1.7.8
The Watts circular dividing machine for graduations on scales up to 54 in diameter (1905).

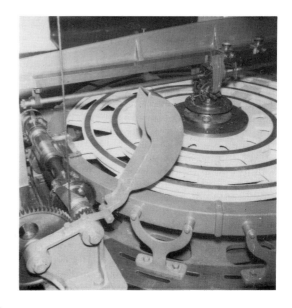

35

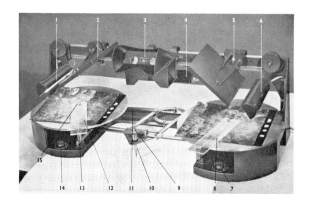

1.7.9

The radial line plotter is a simple, inexpensive instrument for obtaining planimetric details from aerial photographs. It is based on the radial line method and is an ideal instrument for plotting detail from pairs of photographs on which control points have been established by direct ground survey, or from existing maps. It is useful for forestry development and geological survey.

The instrument is used on a drawing board and is speedy in operation and sufficiently portable to be used by mobile survey parties. The photographs are observed through a stereoscope with convenient inclined viewing, and any person with normal stereoscopic vision can be trained to use it in a few hours.

Key: 1, left-hand clamp; 2, left-hand mirror; 3, binocular viewing head; 4, rear lamp; 5, clamp screw for angle setting of mirror; 6, right-hand lamp; 7, ball foot, right-hand; 8, cursor, right-hand; 9, weight for pencil; 10, pencil holder; 11, parallel link mechanism; 12, cursor, left-hand; 13, retaining screw for cursor; 14, ball foot, left-hand; 15, cursor centre pin.

Although the company was established by manufacturing field surveying equipment, an air survey camera, radial line plotters, and Thompson–Watts plotters, developed with the help of Professor E. H. Thompson (previously a Captain in the Royal Engineers and Air Officer to the Ordnance Survey), were made for photogrammetry during the years 1950 to 1970 (Figs 1.7.9 and 1.7.10).

V. W. H. Towns was the Technical Director during development of a substantial part of this range of Watts equipment. The air survey camera, designed for modern photogrammetry, incorporated a register glass provided with a 'reseau' or network of fine marks by which film dimensions are controlled in analytical air triangulation. A 152 mm (6 in) $f/4.5$ lens with a field angle of 93° specially for the camera was made by Wray (Optical Works) Ltd, Bromley, Kent.

On the same factory site at Camberwell, their subsidiary company, James Swift and Son Ltd, which was established in 1853, pioneered in Britain the manufacture of

1.7.10

The Thompson–Watts plotter is a first-order stereoscopic instrument for producing maps to any desired scale from aerial photographs, and is based on principles developed by Professor E. H. Thompson.

Its outstanding feature is that relative orientation is entirely independent of absolute orientation, so that photographs can be set up quickly and easily. Another great advantage of the instrument is that only three mechanical movements are necessary for the reconstruction of the space model. Provision is made for carrying out aerial triangulation, and a practical solution has been found to many problems inherent in this type of work. With the Thompson–Watts plotter, any shape or mark, including circular marks, can be used for correspondence setting.

The optical system of the instrument, which is based on the Porro principle, is compact and has an exceptionally large field of view.

polarising microscopes for the petrologist, mineralogist and chemist. By 1967, Swift microscopes had become unprofitable against large-scale competition and their manufacture was terminated.

Frank Watts died in 1939 and was succeeded by George Watts as Chairman of the Company. George A. Whipple, C.B.E., M.A., C.Eng., F.I.E.E., M.I.Mech.E., joined as Managing Director and, when the Company amalgamated with Adam Hilger Ltd in 1948, he became Chairman of Hilger and Watts Ltd until 1968 when the Company became a subsidiary of Rank Precision Industries Ltd.

Wray (Optical Works) Ltd

The firm of William Wray was founded in 1850 by a solicitor who, like so many gentlemen of his time, made his own telescopes and was an amateur astronomer. Because of his success in polishing objective lenses, he started a business and, when photography was invented, he made a range of portrait lenses.

One of the earliest anastigmat lenses, named the 'Platistigmat', was designed and made by Wray in about 1890, but the barium glass used in the design was unstable and after some years the glass crystallised so making the lens unusable. The business declined, and by 1908 little remained but a few items of stock and some old machinery.

At that time, Albert Arthur Smith, who was manager of the optical firm, Ross of Clapham, was having discussions with Irvine G. Aitchison (later the firm became Dollond and Aitchison) who had set up a business in the Strand, London, to make prism binoculars. It was decided that Mr Smith would leave Ross and purchase the Wray Company which would be combined with the Aitchison Company.

The new company Wray (Optical Works) Ltd continued to manufacture Aitchison prism binoculars, introduced the Wray 'Lustrar' range of photographic lenses, and soon needed more space, so in 1910 the factory was moved to Peckham. During the 1914–18 war, large numbers of binoculars were made and Wray Lustrar lenses were used on enlargers for the Royal Flying Corps. Because of the need for more space, Mr Smith moved the factory to

Bromley in 1915.

After the war, the British Government set up a Technical Optics Department in the Royal College of Science, Imperial College, and among the early students were Arthur William Smith, son of Albert Arthur Smith, R. E. Reason, who later joined Taylor, Taylor and Hobson Ltd, and R. Kingslake, who subsequently became Professor of Optics at Rochester University and Director of Optical Design at Eastman Kodak, Rochester, U.S.A.

From 1924, A. W. Smith worked full-time for Wray and in 1928 he was made a Director so joining his father A. A. Smith and I. G. Aitchison on the Board. The photographic lens business expanded and Wray lenses were fitted to Kodak cameras and Soho Reflex cameras made by Kershaw. Process lenses were developed for graphic arts cameras, made by Hunter Penrose Ltd, and reconnaissance lenses were developed for aerial cameras made by the Williamson Manufacturing Co. Ltd and used by the Royal Air Force.

Mr A. A. Smith, the virtual founder of the company, retired in 1940 when he was 76 years old and the control of the business passed to his son, Arthur William Smith, who became Managing Director.

After the essential production of war supplies in 1939–45, there was a great shortage of all types of optical equipment and the invention of electronic computers enabled much-improved lens systems to be developed. Dr Charles Wynne was responsible for optical design at Wrays for 16 years from 1942 and raised the standard of design, production and testing to a very high level.

When Dr Wynne left Wrays, to join the Royal College of Science as Director of Optical Design, his successor was David Day, M.Sc., who had joined the company in 1950 and was responsible to Dr Wynne.

A substantial export business in high-quality lenses to Sweden, Switzerland, Germany and America was built up during the 1950s and early 1960s, based on the designs of Charles Wynne, but the most important products of the Wray Company from its earliest days were prism binoculars and the Japanese competition in the middle 1960s slowly eroded the market.

In order to replace this lost business, a 35 mm enlarger and slide projector were developed with some considerable success.

An air survey camera designed by Hilger and Watts Ltd, and later developed by Williamsons, was fitted with a 152 mm (6 in) $f/5\cdot6$ Wray lens for use with first-order plotting machines.

In 1962, because there was no successor to own and manage the Wray Company, Arthur Smith welcomed an approach by George Whipple, Chairman of Hilger and Watts Ltd. So Wrays became a subsidiary company of the Hilger and Watts Group and for seven years the two firms worked together very satisfactorily.

By 1968, Hilger and Watts Ltd had reached a desperate financial position and a bid from the Rank Organisation was accepted by the shareholders in July of that year. As part of the reorganisation, and attempts to reduce expenditure, the Wray factory at Bromley closed in 1971 some six months after the retirement of Arthur Smith.

W. Ottway and Co. Ltd

For nearly 330 years from 1640, the Ottway firm and company was passed down from father to son in unbroken succession. The founder was Francis Ottway and the last Managing Director was Geoffrey C. Ottway. The Ottway family Bible records the baptism of Francis Ottway's daughter Rebecca in February 1641 (Fig. 1.7.11).

In 1640, the Ottway firm was established at the Royal Exchange, London, but on 2 September 1666 a baker's shop in Pudding Lane, near London Bridge, caught fire and the flames spread along the riverside wharves. By the second day, buildings near the Royal Exchange, in Lombard Street, and in the Cornhill were ablaze. The Great Fire of London destroyed the Ottway premises, stock and records in what was described by Samuel Pepys in his *Diary* (1668) as a 'hugh bow of flame'.

The firm was restarted in Cary Street, Strand, where it remained until 1700 when new premises were acquired in King Street, Holborn. For the next 100 years, the firm settled down to make telescopes and other optical instruments along with contemporary opticians and engineers such as Chester Moor Hall, John Dollond, Peter Dollond, James Short, Sir William Hershel and Jesse Ramsden (section 2.1).

In 1800, the firm moved to Clerkenwell and during the nineteenth and early twentieth centuries a wide range of surveying instruments, and astronomical and sighting telescopes, were

1.7.11
Family Bible showing birth records of seventeenth-century Ottway children.

developed. For expansion of manufacture, another move was made in 1899 to extensive works in Ealing.

In 1893, William Close Ottway, grandfather of Geoffrey C. Ottway, in collaboration with a Mr Crofts and anticipating the Lumière brothers, invented a cinematograph projector using Kodak unperforated film $2\frac{1}{2}$ in wide. During the early part of the twentieth century, the Ottway Company made many motion-picture projectors incorporating a Maltese cross developed by Robert W. Paul in 1896 (Fig. 1.7.12).

During the period of the Boer War, and leading up to the 1914–18 Great War, production was concentrated on naval gunsighting telescopes (1902), variable-power spotting

1.7.13
Heliographs as supplied to the War Office by W. Ottway & Co. Ltd.

telescopes (1906), look-out and long-distance telescopes for the officers-of-the-watch and coastguards, and also binoculars and heliographs which were extensively used for communication in India (Fig. 1.7.13).

The firm became a private company in 1911, and when William Close Ottway died in 1914 the organisation continued under three brothers. Geoffrey C. Ottway, son of C. E. Ottway, joined the company in 1930 and later became Managing Director.

In 1947, W. Ottway and Co. Ltd became a public company but, due to intense competition and the need to develop a new range of equipment, the company merged with Hilger and Watts Ltd in 1961. When Hilger and Watts Ltd was merged with Rank Precision Industries Ltd in 1968, Geoffrey C. Ottway retired and the Ealing factory was closed.

1.7.12
Early cinematograph designed about 1893 using unperforated roll film 2·5 in wide.

The Worshipful Company of Spectacle Makers was granted a Royal Charter in 1629. One of the primary purposes of that ancient guild was to maintain good standards of workmanship in the making of spectacles offered for sale in London.

There was a decline in the industry during the nineteenth century and by the beginning of the twentieth century the larger part of the domestic demand for spectacles was met by imports from North America, France and Germany.

The need for a strong optical industry was brought home during World War I and in June 1919 the United Kingdom Optical Company Ltd was formed as a private company by Mr F. W. Watson-Baker (of the old established optical firm W. Watson and Sons Ltd), Messrs G. W. Bayliss, E. Culver, S. Culver and A. H. Emerson. Production was the responsibility of Edward Culver and Alfred H. Emerson (1872–1947) who became joint Managing Directors.

Some of the original plant came from the factories of George Culver Ltd and H. Emerson and Son which had been formed by A. H. Emerson and his father before the Boer War.

Alfred H. Emerson invented the one-piece bifocal lens and in 1906 he formed the Uni-Bifocal Company in order to manufacture solid bifocal lenses which he also licensed to be made in France, Germany and America. At that time, J. A. Moore was one of his apprentices.

Solid bifocals are produced by grinding different curvatures on the distance and reading portions of a bifocal surface. A prism-segment solid bifocal incorporates a prism in the reading portion of a lens and the blank is tilted during grinding of the segment. A centre-controlled solid bifocal has a distance lens into which a segment surface has been ground, resulting in a step extending along the dividing line (Fig. 1.8.1).

Production at U.K.O. began in September 1919 and by 1924 all preliminary expenses incurred by the firm had been written off and the first dividend paid. A notable improvement in fused bifocal lens design was first suggested by a London optician, Mr F. Bruce Watson, and the now world-famous Univis fused bifocal was developed in the U.K.O. Mill Hill, London, factory and was soon being exported to America and Europe. The segment of standard fused bifocal lenses was round, but the patented Univis fused bifocal was made in a variety of shaped segments of which types B and D became very popular (Fig. 1.8.2).

A 32 mm depression was worked on the crown glass by methods as for round segments. The shape of the final segment was produced by a composite flint button consisting of two or

1.8.1
(a) Prism-segment solid bifocal lens. (b) Centre-controlled solid bifocal lens.

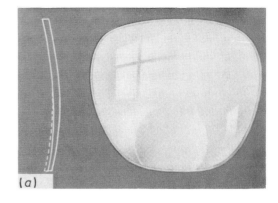

(a)

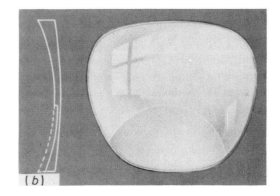

(b)

more pieces of glass. The B shape button consisted of a narrow rectangular moulding of flint glass, which was ground and smoothed on the two longer edges, and two similar shaped pieces of crown glass ground and smoothed on one edge. These three pieces of glass were carefully cleaned and clamped together with the flint between the crowns and the smoothed surfaces in contact. In this condition, it was passed through the lehr and fused into a solid piece of glass. The D shape was flint with a rectangular strip of crown glass. When the button had been fused, it was ground and polished to the required curve in the usual way.

In 1926, there was an agreement to manufacture Univis bifocals under licence in the U.S.A. and this helped to establish the Univis Lens Company at Dayton, Ohio. Stanley A. Emerson, son of Alfred H. Emerson, later became Vice President of Research and Development after having assisted in setting up this factory in America.

By 1935, new buildings were needed at the Mill Hill factory of U.K.O. and refinancing led to the formation of a public company in 1936. Mr F. W. Watson-Baker remained as Chairman but E. C. Sherren replaced A. H. Emerson as a director and Dr Clay was invited to join the Board. Messrs Taylor and Ustonson Ltd then amalgamated with U.K.O. bringing to the Board C. B. Ustonson, as well as J. A. Moore as a joint Managing Director with E. C. Sherren and E. Culver.

In 1938, J. A. Moore visited Australia, New Zealand, U.S.A. and Canada. When F. W. Watson-Baker resigned as Chairman in 1939, his place was taken by E. Culver with J. A. Moore as Deputy Chairman. During World War II, precision optical components were made under the supervision of A. A. S. Moore, son of J. A. Moore, who later became Technical Director of U.K.O.

In 1946, a new factory was started in Lurgan, Northern Ireland, for the manufacture of spectacle frames as well as ophthalmic lenses. New finance, and amalgamation with several other companies, followed in 1949 and E. C. Sherren became Chairman of the first Holding Company Board. When Mr Sherren died in 1951, J. A. Moore became Chairman of the United Kingdom Optical Company Ltd and E. E. Snow was elected Chairman of the Holding Company.

In 1954, The Emerson Optical and Manufacturing Company Ltd was acquired, which had been founded in 1936 (by A. H. Emerson when he retired from U.K.O. and his son Stanley A. Emerson) for the manufacture of precision lenses and prisms.

Over the period of more than 50 years since

1.8.2
Fused bifocal lenses. Univis type B segment is approximately 22 mm × 9 mm and the shape enables objects to be seen through the distance portion under the segments. Univis type D segment is approximately 22 mm × 16 mm.

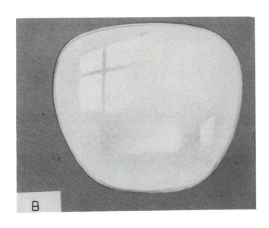

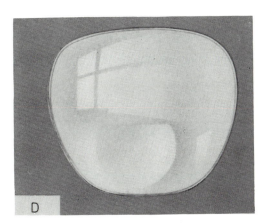

1919, the United Kingdom Optical Company has developed to become the largest manufacturer of mass-produced spectacle lenses and frames in the British Isles with factories in England, Scotland, Ireland and Wales. Because the first Chairman was F. W. Watson-Baker, it is of interest to give a few details of the Watson company which is now owned by Philips Electrical and Associated Companies Ltd (section 1.10).

1.9 Vickers Instruments

Vickers Ltd registered the name of Cooke, Troughton and Simms Ltd, York, in 1922, which combined T. Cooke and Sons Ltd, controlled by Vickers Ltd from 1915, with Troughton and Simms Ltd, London. In 1959, the company acquired the share capital of C. Baker of Holborn Ltd, a well known firm of optical instrument makers, which was established in 1765.

Troughton and Simms Ltd

John Rowley (1673–1728) is known to have been established in business on his own account prior to 1700 because an Analemmatic Dial, bearing his name, is dated 1698. Many instruments, made by Rowley between 1698 and 1720, are still in existence, particularly in the collection which Charles Boyle, Fourth Earl of Orrery (1676–1731), bequeathed to Christ Church, Oxford. These include drawing instruments, pocket dials, proportional compasses, parallel rulers, protractors, artillery scales, quadrants, globes, planetaria, levels, sextants, the 36 in Standard of Length and the celebrated 'Orreries'. A 300 mm telescope, with cross wires and a 150 mm level vial, belonged to Sir Isaac Newton (1643–1727) and is marked 1703, the year in which he became President of the Royal Society.

Attempts had been made to reproduce mechanically the movements of the solar system, and simple clockwork models reproduced the path of the Moon round the Earth. In 1700, George Graham (1673–1751), an eminent clock- and watchmaker who understood practical astronomy, invented a working model of the whole Copernican planetary system for the Fourth Earl of Orrery and named it an 'Orrery'. John Rowley improved the design, and was responsible for the addition of the in-

ferior planets; Thomas Wright is credited with the superior planets. John Rowley probably retired from instrument making about 1715 after receiving his Royal Appointment as Master of the Mechanicks to George I.

Thomas Wright succeeded to the business of John Rowley in London and, after 1718, openly traded under the sign of 'The Orrery and Globe' which he acquired the right to use from Charles Boyle, the Fourth Earl of Orrery. In 1727, Thomas Wright was appointed Mathematical Instrument Maker in Ordinary to King George II. In 1731, he was described as a 'Perspective and Optic Glass Maker' and, although globes and orreries were the principal products, a wide range of other types of instrument were made, including magnetic compasses and compound microscopes. From 1720, George Graham and Thomas Wright occupied the same building in Fleet Street where Wright had his instrument shop until he retired in favour of Benjamin Cole during 1748.

Benjamin Cole, Sr (1695–1766), entered the service of Thomas Wright at the sign of 'The Orrery and Globe', but at some time between 1733 and 1739 started a business on his own account in Fleet Street. In 1739 his son, also Benjamin, was apprenticed to him, which indicates that he was no longer a servant of Thomas Wright.

After taking over the business of Thomas Wright in 1748, he traded under the name B. Cole and later Cole and Son at 'The Sign of the Orrery'. Orrerys, plane tables for surveying, theodolites, quadrants, octants and microscopes marked 'B. Cole No. 136 Fleet Street' are in the Science Museum, South Kensington (Fig. 1.9.1).

Benjamin Cole Jr (1725–1813) was elected to the Livery of the Worshipful Company of

1.9.1
An orrery, by Cole (c. 1750). (Reproduced from *Early Science in Oxford*, vol. II, by permission of the executors of the late Dr R. T. Gunther.)

Merchant Tailors in 1763 and succeeded to the business on the death of his father in 1766. Benjamin Cole had negotiations with John Troughton, from 1776 to 1781, who moved to No. 136 Fleet Street, London, in 1782 to take over the business in partnership with his brother Edward Troughton.

John Troughton was one of a number of London Instrument Makers who, in 1764, signed an unsuccessful petition to the Privy Council for the revocation of John Dollond's patent on the achromatic objective lens. He also completed a dividing engine which 'at the full stretch of his pecuniary means' had occupied him for three years. After the death of John Troughton in 1784, the business was continued by Edward Troughton (1753–1836) who was sole proprietor for 47 years.

Edward Troughton's most important contribution to the development of astronomical instruments was with improvements in the methods of dividing a circle. He claimed 'to have reduced the time of originally dividing a 4 foot meridian circle from an average of 52 days to 13 days and in addition to have improved the accuracy'.

In 1793, Troughton completed a circular dividing engine which permitted him to cut 24 divisions a minute. Many important instruments are in existence bearing the name of Troughton, including his Imperial Standards of Length and 24 in transit theodolites supplied for the Trigonometrical Surveys by the Board of Ordnance.

Through the quality of his products, Troughton gained a reputation as the foremost maker of sextants and finely divided mural circles. In 1809, he was awarded the Copley medal of the Royal Society, and a year later became a Fellow of the Royal Society.

In 1826, he merged his business with that of William Simms (1793–1860), under the name Troughton and Simms, but in 1831 Edward Troughton retired from active business and lived with the family of William Simms until his death in 1836.

It is interesting to note that Joseph Beck, one of the founders of the firm R. and J. Beck in 1867, was an articled apprentice with Troughton and Simms (section 1.5).

William Simms made improvements to Troughton's dividing engine and was responsible for an orginal self-acting mechanism. In 1843, a new dividing machine was completed and 'he reduced the process of graduation of instruments from a work of weeks to a work of hours by the invention of his self-acting dividing machine'. During the years 1845 to 1860, a large number of high-quality theodolites and levels were made and some of these are now on display in the museum at the Ordnance Survey headquarters at Southampton. In 1852,

William Simms was elected a Fellow of the Royal Society.

After the death of William Simms in 1860, his second son James Simms (1828–1915) entered into a deed of partnership with his cousin William (1817–1907), who had worked the dividing machine to produce many theodolite and transit circles.

In 1871, William retired and James was sole proprietor until his death in 1915. The business was continued by the two sons of James Simms, who were William (1860–1938) and James (1862–1939), but incorporated as Troughton and Simms Ltd. In 1922, the Company was amalgamated with T. Cooke and Sons Ltd of York and named Cooke, Troughton and Simms Ltd.

T. Cooke and Sons Ltd

Thomas Cooke (1807–68) was the founder of T. Cooke and Sons and in 1837 set up for himself as a manufacturer of reflecting and refracting telescopes. In the report of the British Association for 1853, Professor Phillips refers to an 'achromatic objective of $6\frac{1}{4}$ inches aperture, the work of our excellent artist Cooke, which is driven equatorially by a very equable clock movement'.

Cooke was one of the first to apply the principles of the mechanical engineer to the construction of optical instruments and so achieved a considerable output of equatorial telescopes, of eight to ten inches aperture, and many smaller instruments. After moving to their new Buckingham Works at York in 1855, Cooke, with the assistance of his two sons, Frederick and Thomas, commenced making surveying instruments, turret clocks and sidereal clocks for the astronomer. About this time, turning and screw-cutting lathes, wheel-cutting machines and dividing engines were also developed (Fig. 1.9.2).

After the death of Thomas Cooke in 1868, the business was managed by his two sons until 1894, when Frederick retired in favour of Alfred Taylor (1863–1940), and in 1897 the firm became the company T. Cooke and Sons Ltd.

At that time, H. Dennis Taylor (1861–1943) was manager of the optical workshops and had

1.9.2
Buckingham Works rules, 1865.

invented the first apochromatic astronomical object glass in 1892 followed by the 'Cooke' photographic lens in 1894 which was to play so large a part in the future development of photography. The manufacturing rights for the 'Cooke' photographic lens were sold to Taylor, Taylor and Hobson Ltd in 1893.

About this time, Dennis Taylor invented a new form of star test for telescope objectives, known as the method of autocollimation, which provided a test object at infinity in the shape of an artificial star.

Between the years 1902 and 1920, Dennis Taylor devoted himself to the improvement of naval rangefinders and introduced the 'built-up' end reflectors, the swinging range prism, and binocular vision. In 1906, he published his *System of Applied Optics* and for his achievements was awarded the Trail Taylor Medal of the Royal Photographic Society and the Duddell Medal of the Physical Society of London.

T. Cooke and Sons Ltd was acquired by Vickers Ltd in 1916, and Troughton and Simms Ltd was purchased some years later. In 1922, the businesses were merged and the name was changed to Cooke, Troughton and Simms Ltd.

In 1923, the Carl Zeiss Company of Jena introduced a theodolite, designed by Dr Heinrich Wild, in which the circles were of glass with etched figures and graduations. The graduations were read with an optical micrometer whereby diametrically opposite graduations were meaned to the nearest second of arc. This degree of accuracy was previously unobtainable from graduations on metal scales made of silver, gold or platinum (section 8.3).

The Admiralty called a meeting at Tavistock to consider the implications of this development in Jena, and an alternative method of obtaining a meaned reading was discovered using a 'light gap' instead of a 'coincidence setting'. The Cooke Tavistock theodolite was the first double-reading optical micrometer theodolite to be made in Britain and has been in production since 1927 (Fig. 1.9.3).

1.9.3
(a) Cooke Tavistock theodolite reading to one second of arc. (b) Cooke Tavistock theodolite. The horizontal and vertical glass circles are 85 and 70 mm diameter, respectively.

(a)

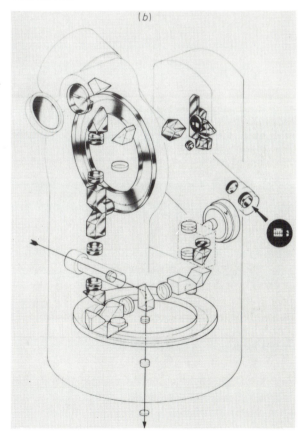

(b)

When Sir Frank Dyson was Astronomer Royal in 1935, the Admiralty placed an order for a Reversible Transit to have 24 in glass circles and a 7 in objective lens. This instrument, sometimes called a Meridian Circle, consists of a telescope supported between trunnions on an east–west line and as the telescope moves it is confined to the meridian. Star passages across the meridian, as the Earth rotates, are timed each day at the Royal Observatory, Herstmonceux (Fig. 1.9.4).

This was the last of a very long line of astronomical instruments made by the Company as, in 1938, the goodwill of the astronomical business was sold to Sir Howard Grubb, Parsons and Co. Ltd, of Newcastle upon Tyne.

1.9.4
Observing star positions with the Meridian Circle telescope made by Cooke, Troughton and Simms Ltd. This transit telescope uses a circular scale for measurement and a pool of mercury for calibration.

1.10 Watson

The firm of William Watson was founded in 1837, and by 1875 it was manufacturing cameras, lenses, magic lanterns and microscopes. During the next 10 years surveying instruments and astronomical and portable telescopes were added to the sales catalogue. By 1906, the firm had become W. Watson and Sons and, in 1908, W. Watson and Sons Ltd was formed with a factory at Barnet. In 1919, F. W. Watson-Baker became Chairman of a new private company called The United Kingdom Optical Company Ltd (section 1.8).

It is interesting to note that early in this century Marconi was a frequent visitor to Watsons at Barnet who made for him induction coils and other apparatus for his wireless experiments. Also, during 1931–32 Watsons made the 'Fultograph' which was used by the *Daily Mail* for the transmission of photographs by wire between London and Manchester. All family connections with the firm ended when W. E. Watson-Baker sold his interest in 1949.

Watson microscopes, introduced about 1874, were described in the tenth edition (1883) of that classic book on microscopy *The Microscope* by Jabez Hogg and the range of Watson microscopes continued to be important products until 1970.

2 Telescopes

2.1 Introduction

The eye is a 'thick lens' which has a fixed image distance and is capable of varying its focal length. An image is formed on the sensitive retina at the back of the eye and, as the object moves, so muscles squeeze the crystalline lens and change the radius of curvature of each surface (Fig. 2.1.1).

If white light passes through a single convex lens, the image formed is fuzzy and coloured. This coloration is due to chromatic aberrations that arise because white light is a mixture of all colours. When passing through a transparent substance, blue light is bent more than red light, the amount depending on the refractive index of the substance. Chromatic aberration in a lens can be in the form of lateral or longitudinal aberrations.

Lateral chromatic aberrations are formed by the lens of a human eye and are independent of the size of the iris. Because red and blue rays which pass through the centre of the lens change direction due to refraction, then the inverted images formed on the retina will be of different sizes and colours. Lateral colour in a lens becomes more intense as the image point moves from the centre of the field, and it is proportional to the distance of the image point from the centre. Lateral colour is not altered by changing the aperture of the lens. The blue edge of a light patch may be closer to the lens axis than the red edge, as with a human eye, or, depending on the lens construction, it may be the other way round (Fig. 2.1.2).

Longitudinal chromatic aberrations increase

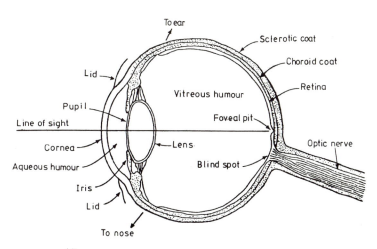

2.1.1
Human eye. When the eye concentrates on an object, the object's image is focused on the foveal pit of the retina.

47

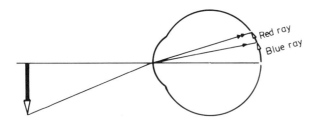

2.1.2
Lateral chromatic aberration in a human eye.

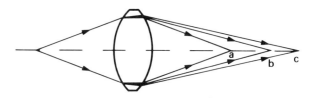

2.1.3
Longitudinal chromatic aberration. The white rays from a point object are broken up inside the lens and come to foci that depend for their position on wavelength. The blue rays focus at a, the yellow-green at b, and the red at c. If a screen were placed at b, the image formed on it would be surrounded by a purple halo due to the circles of confusion formed by the red and blue rays.

in proportion to the aperture of a lens, and the 'bending power' of the refracting substance is not the same for all colours of light. This variation is known as the dispersion power (Fig. 2.1.3).

Although the human eye is uncorrected for both lateral and longitudinal chromatic aberrations, it automatically accommodates so as to bring the brightest part of the spectrum into focus on the retina. The blue then focuses in front and the red behind the retina. The halo about the image point, which is formed by the blue and red rays, appears purple, and is unseen in daylight because the iris aperture in front of the crystalline lens is very small (Fig. 2.1.4).

The retina of the eye consists of a mosaic of two types of receptor cell, known as rods and cones. The rods are more sensitive to light but cannot distinguish between colours, and they occur mostly towards the outer areas of the retina. The cones provide colour vision and are concentrated near the fovea, at the centre of the retina, which is the high-resolution zone.

The cones contain three photosensitive pigments and whenever the red and green components of an image are in focus, at the fovea, the blue component is out of focus due to chromatic aberrations. This effect is amplified by the separation of the blue pigment absorption curve from those for the red and green, as Nature has decreased the importance of blue light for high-sensitivity night vision (Fig. 2.1.5).

Night myopia was first discussed by Lord Rayleigh in 1883, when it was found that the

2.1.4
Under normal circumstances, the eye automatically accommodates to bring the brightest part of the spectrum into focus on the retina. The blue then focuses in front of the retina, the red behind it. The halo about the image point formed by the blue and red rays is purple. It is not ordinarily seen but, by covering part of the entrance pupil, it can be brought into view with a point source of light.

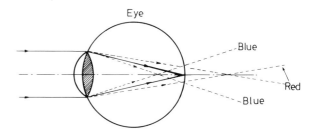

48

eye has about 2 dioptres more power for blue light than for red light. Because of the Purkinje chromatic shift, we are more short-sighted in low light intensities and there should be provision for this aberration in the design of night telescope eyepieces.

The inverted image on the retina is automatically erected by the brain and this normal condition must be provided by a magnifying optical system. Astronomical telescopes usually do not provide an image-erecting system because it does not matter in which way a star is viewed. Some surveying telescopes, for simplicity of design, do not provide an image-erecting system and so the figures on staves are sometimes printed inverted for the convenience of surveyors who wish to see erect figures.

It is probable that the first lens telescopes were made in Holland during the early years of the seventeenth century, as in 1608 the optician Hans Lippershey (1587–1619) applied for a patent on his telescope design. In the same year, a double telescope was designed as two afocally combined lenses with negative eyepieces.

In 1609, Galileo Galilei (1564–1642) heard of Lippershey's invention and made a telescope for himself, with a magnification of 3×, consisting of a plano–convex lens as objective and a plano–concave lens as eyepiece. This telescope was used for

astronomical observations, and further developments in telescope design by him led to magnifications of 8× and 32×.

Only a few of the early telescopes have survived, because astronomical telescopes, at that time, were made from cardboard tubes with the top layer covered in parchment. The lenses were fixed in wood or ivory cells and diaphragms prevented internal reflections. The single-lens objectives, before the introduction of the achromatic lens by John Dollond in 1759, were satisfactory only if the telescope tube was very long or the magnification was below 3×.

Thin brass telescope tubes were introduced around 1780 by Peter Dollond and were used for over 150 years. Brass tubes added to the precision of the telescopes, made them smaller when folded, and avoided damage by salt water when they were used at sea.

The Galilean type of telescope, with a positive objective and negative eyepiece, has an erect virtual image and small field of view, but is no longer used except for cheap field or opera glasses. As no real image is formed, there is no place for a graticule and it is not possible to achieve a magnification of more than two or three powers. The short structural length (d) is equal to the difference between the absolute focal lengths of the objective (f_o) and eyepiece (f_e) (Fig. 2.1.6).

In 1611, Johann Kepler published his *Dioptrice* and described the detailed operation of a Galilean telescope with concave eyepiece and his own proposal for a telescope with a convex objective and a convex eyepiece (Fig. 2.1.7).

2.1.5
The eye's red and green pigment absorption curves are close together.

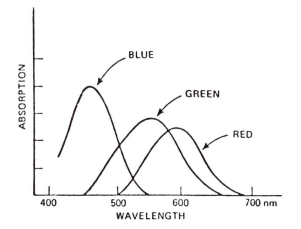

2.1.6
The Galilean telescope with erect image. A Huygen's eyepiece may be used instead of the concave lens.

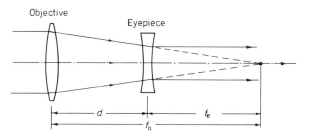

49

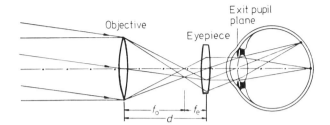

2.1.7
Keplerian astronomical telescope with infinite conjugates and real inverted image in front of the eyepiece.

This telescope, with two convex lenses, gives an inverted representation of a distant point but in a three-lens compound instrument, such as his terrestrial telescope, Kepler was able to erect the inverted real image. The Keplerian telescope has a relatively large field of view with a real image, by which objects examined through it can be measured. A terrestrial ship's telescope made by Peter Dollond in 1783 used a cemented doublet objective, a brass draw tube, and a Kellner eyepiece. Telescope objectives usually cover a semi-field of only 2 or 3° so, if most of the chromatic and spherical aberrations are corrected, the outstanding astigmatism and curvature of field can usually be tolerated (Fig. 2.1.8).

2.1.8
(a) Doublet telescope objective corrected for longitudinal chromatic aberration. (b) Terrestrial refracting telescope with erect image and Kellner eyepiece.

Binocular Galilean telescopes were developed in order to eliminate the need for a long draw tube. By fitting achromatic objectives, a much larger field of view was obtained, and from the end of the eighteenth century these small but powerful instruments were on sale.

The first prism binoculars were designed by Ernst Abbe and the firm of Carl Zeiss Jena obtained a patent in 1894. In 1910, opera glasses, with image reversal by double Porro prisms, were produced for the first time in Jena. Their main features, compared to the traditional Galilean system, are the larger field of vision and smaller and lighter-weight objectives. A sharp outline of the field of vision is given by a diaphragm in the eyepiece image plane, and image erection is achieved by the prisms without additional erecting lenses (Fig. 2.1.9).

Rangefinders have been developed from prism binoculars and are used for field surveying and military purposes. In order to calculate distance with a coincidence

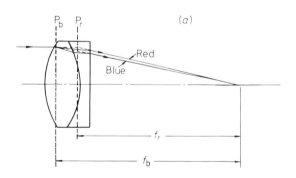

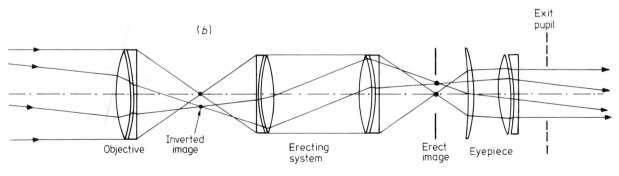

50

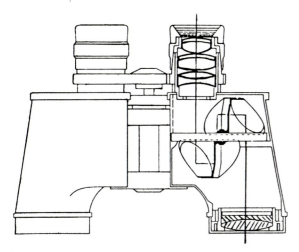

2.1.9
Principle of refracting binoculars based on Galilean telescope, using double Porro prisms for erection of image and a Kellner-type eyepiece.

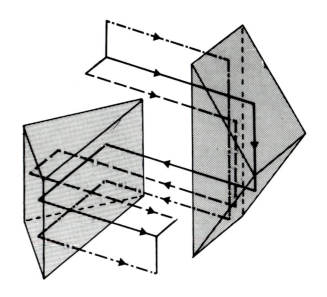

rangefinder, it is necessary to know the length of one side and one acute angle of a right-angled triangle. The base b of a rangefinder may be 1 or 2 m or longer and have objective lenses of focal length f'. If the target T is at a finite distance R, then light from the target will arrive along lines TA and TB (Fig. 2.1.10).

The displacement d, equal to ED, is known as the parallactic displacement. If prisms are placed at A and B to deviate the beams TA and TB through 90°, and the two objectives f' will form images in a common plane, then the upper half of the image formed by one objective and the lower half of the other will be reflected by the two prisms P into the eyepiece where they can be brought into coincidence by the adjustable prism at G. As the prism moves along the optical axis, the deviation changes until the two images are in coincidence. The scale can be calibrated in units of distance (Fig. 2.1.11).

Stereoscopic rangefinders are based on the same principles as are used in aerial survey and photogrammetry. Two objects at different distances from the eye can be stereoscopically resolved if the angles subtended by the eyes of

2.1.10
The displacement d, equal to ED, is called the parallactic displacement and is seen to be a measure of the range, $R = bf'/d$. This is the fundamental equation for the coincidence rangefinder and also holds for the stereoscopic instrument.

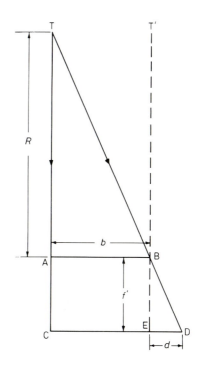

51

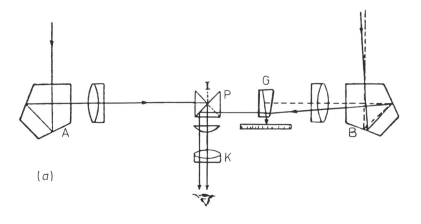

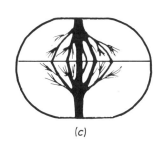

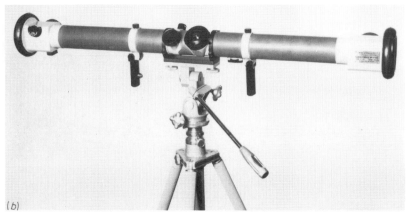

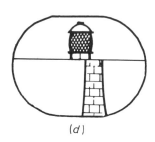

2.1.11
(a) Coincidence rangefinder. As the prism G is moved along the optical axis, the rays strike it at different heights and, consequently, different amounts of deviation are produced. Thus, by shifting the prism, the deviation may be changed. (b)\Barr and Stroud Coincidence Rangefinder for ranges between 250 and 10 000 m. The rangefinders utilise the coincidence principle in which the field of view is divided by a horizontal line into two portions. The image of the object being ranged (right eyepiece) is split along this line and, by adjusting the range control, the lower image is shifted horizontally until the image portions are coincident at the line (see (c) and (d)). The range is then read off the direct-reading range scale (left eyepiece). (c) In FT37 (LAND), the image in the lower portion of the field is erect, but that in the upper portion is inverted and is a duplicate of the top section of the erect lower image. This arrangement is advantageous for land applications where targets may be rocks or trees which are of irregular shape and are usually not well defined. (d) In FT37 (MARINE), the image in both portions is erect, as targets are usually well defined. The inverted-image arrangement is not suitable for marine applications since the image would be separated at the horizontal line each time the vessel pitched and rolled.

the observer and the two objects differ by more than 20 to 30 seconds of arc, depending on the eye separation, at a distance of not more than about 700 m. In order to resolve longer distances, an artificially increased interocular distance must be provided (section 7.2).

If two telescopes are mounted to a fixed base, with interocular distances to suit an observer, and graticules with a single dot are placed in the plane of images A and B, then, if one dot is fixed and the other movable, the two can be fused together stereoscopically so that only one dot is suspended in space at the same time as seeing one image of the target (Fig. 2.1.12).

If the movable dot is at D, and the fixed dot at C, then the observer will fuse the two images and see a dot suspended in space at E. If the dot is moved from D to F, then it will appear in space at G so, by moving a graticule, the dot in space will appear to advance or retreat. By this means, the observer can make the dot in space stop over any object he can see in the telescope and, if the graticule mechanism is calibrated, the distance of the object can be measured. Alternatively, if a series of dots are spaced in pairs on the graticule, then the dots in space will appear at different distances and the range of an object can be rapidly determined. In practice, a single floating mark is used, the graticules are fixed, and one of the real images is shifted laterally to achieve the same result (Fig. 2.1.13).

Rotating wedges at E are used to vary the angle between the two beams entering the eyes until they equal the angle represented by a graticule which appears as far away as the object. The rotating wedges consist of two prisms of equal refracting angles which can be rotated about their own planes in opposite directions. This type of deviation mechanism can be located in front of the objective lens so it is not connected to the sensitive optical bar.

The difficulties in manufacturing large-aperture refracting astronomical telescopes are emphasised when objectives of 4 m diameter or more are required by astronomers who need the highest possible resolution from distant star images. A lens must be transparent and free of bubbles. It can only be supported by the rim so it will sag under its own weight and therefore give a different performance at various attitudes. Chromatic errors, due to refraction,

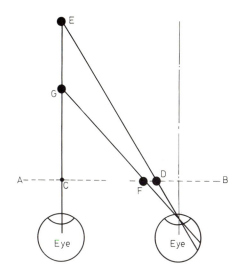

2.1.12
If the movable dot is shifted to F, it is evident that the dot in space will be seen at G. Thus, by laterally moving one of the reticles, the dot seen in space will advance or retreat. By having the mechanism that moves the movable dot calibrated in terms of distance, the observer can read directly the range of any object over which he has stopped the mark.

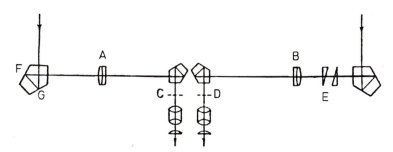

2.1.13
The stereoscopic rangefinder resembles, externally, the coincidence-type instrument, the only difference being its possession of two eyepieces. However, it will be seen that internally there is no connection between the two halves of the instrument, which are essentially only telescopes. The images are fused by purely psychological means.

53

must be corrected in the lens system.

Most modern telescopes are therefore fitted with mirror objectives which need not be made of transparent material and can be supported across the back as well as from the rim. Because there is no refraction in a mirror there are no chromatic aberrations. This is of importance for spectroscopy and any work involving the quantitative or qualitative analysis of light.

The reflecting telescope was invented by James Gregory (1638–75) in 1661. A reflecting telescope consists of a concave paraboloidal mirror coated with a thin aluminium film which accepts parallel light from a distant star and brings it to a focus where the image can be examined, spectroscopically analysed, or photographically recorded.

If a mirror is used as the objective, then the image must be formed in a convenient position outside of the light beam and three systems have been developed to achieve this: these are known as Newtonian, Gregorian and Cassegrain (Fig. 2.1.14).

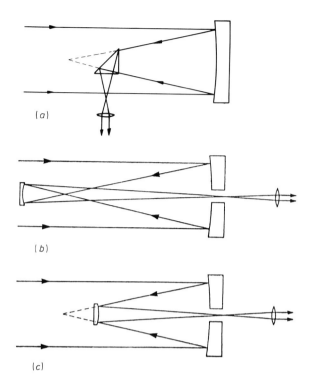

2.1.14
Optical systems of (a) Newtonian, (b) Gregorian and (c) Cassegrain reflecting telescopes.

2.1.15
The three principal types of reflecting telescope. F' is the Cassegrain focus.

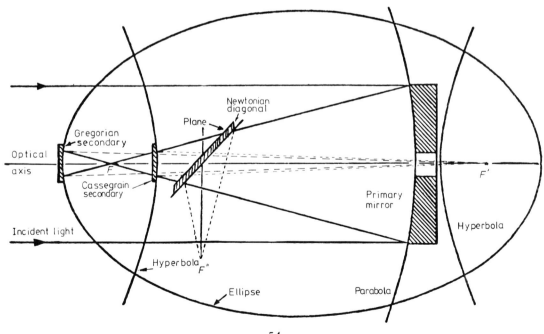

54

The Newtonian system includes a plane mirror or totally reflecting prism to divert the rays to one side of the telescope tube. The Gregorian system includes a small concave ellipsoidal mirror to reflect the rays through a hole in the centre of the primary mirror to form an inverted image behind it. The Cassegrain system uses a small convex hyperbolic secondary mirror to reflect the rays through a hole in the primary mirror (Fig. 2.1.15).

The Newtonian and Cassegrain systems are the most popular types of reflecting telescope, and some are designed to be dual-purpose depending on the performance and focal length required by the astronomer.

2.2 Astronomical telescopes

The ability of a telescope to detect faint stars depends on the aperture of the objective, the quality of optical elements and the focal length of the system.

Amateur astronomers will often use Newtonian reflecting telescopes of about 1000 mm (40 in) focal length and apertures of 100 mm (4 in) to 150 mm (6 in) which will resolve 1·4 to 0·9 seconds of arc, respectively. Serious amateurs will require a Cassegrain telescope with an aperture of at least 250 mm (10 in) mounted on an equatorial head with motor drive and setting circles. Telescopes of this type are made by many optical companies, but only a few have the complex machinery, skills and experience to manufacture large telescopes for astronomical observatories.

Carl Zeiss

Carl Zeiss (1816–88) established his optical workshop in Jena during 1846 and was joined by Ernst Abbe (1840–1905) as a scientific assistant in 1866. In 1889, Abbe founded the Carl Zeiss Stiftung which purchased the firm of Carl Zeiss in 1891.

V.E.B. Carl Zeiss Jena are one of the optical companies with large astronomical telescopes as a traditional product line. Because of the small numbers of these expensive instruments, the design has usually been influenced by the individual wishes of astronomers and standardisation has been difficult to achieve.

The concept of a Universal Reflecting Telescope was intended to standardise production and the Schmidt camera variant has been particularly successful with a

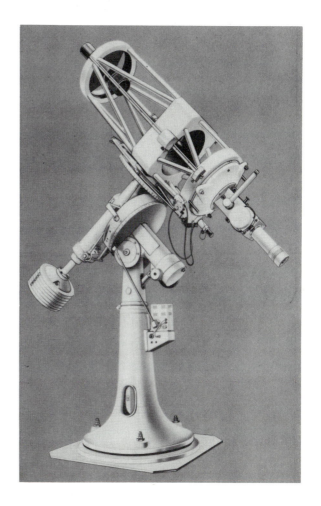

2.2.1
V.E.B. Carl Zeiss Jena Cassegrain mirror system with a free aperture of 600 mm and an equivalent focal length of 7500 mm. The outer diameter of the primary mirror is 630 mm, the thickness at the edge is 110 mm and the focal length is 2400 mm.

2.2.2
(*a*) Universal Astronomical Grating
Spectrograph as fitted to the Cassegrain
Reflecting Telescope 600. The Schmidt
mirror system has an aperture of 125 mm
and focal lengths of 110 mm and 175 mm.
The plate size is 11·5 mm × 50 mm and
spectrum range 330 nm to 900 nm
wavelength. (*b*) Schematic diagram.

(*b*)

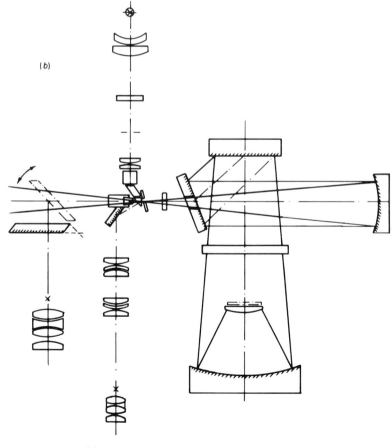

superior performance and resolving power. The technology was developed with a prototype reflecting telescope of 600 mm diameter which has Cassegrain and Coudé variants. The field of view in the Cassegrain focus is about 20 minutes of arc. A synchronous motor for guiding provides automatic compensation for the Earth's diurnal rotation (Fig. 2.2.1).

Spectrographs are required by astronomers who are exploring physical conditions in the universe. A Universal Astronomical Grating Spectrograph was designed to suit the Cassegrain Reflecting Telescope 600. The use of optical mirrors, reflection gratings and ultraviolet transmitting materials for refracting optical elements gives a luminous intensity suitable for spectrographic emulsions and high-speed cameras. One reference line spectrum each is generated for line identification on either side of the stellar spectrum. The light from a reference source, iron arc or neon tube, is easily interchangeable and imaged to the slit through quartz optical elements. Deflection in the collimator direction is controlled by two quartz prisms arranged in front of the slit (Fig. 2.2.2).

When setting for the object to be examined, a flat mirror must be deflected so that stellar light illuminates the graticule of a field-observing eyepiece which should be adjusted with respect to the middle of a slit. When the mirror has been tilted away from the eyepiece, the object can be observed on the slit.

A perforated flat mirror deflects the parallel beam to a diffraction grating which is located in a quick-change mount on the grating table. Six different flat reflection gratings are provided in mounts ready for analysing different wavelengths.

The 1 m Cassegrain–Coudé Reflector Telescope is a variant of the Universal Telescope and has an English mount with switch cabinet and control desk. Air within the telescope tube may be heated to prevent condensation and dry air is fed to the interior. A motor-driven flap shutter protects the objective mirror (Fig. 2.2.3).

2.2.3
The V.E.B. Carl Zeiss Jena 1 m Cassegrain–Coudé Reflector Telescope on English mounting.

Another variant of the Zeiss Universal Telescope is an automatic Schmidt camera for astrophotography (Fig. 2.2.4).

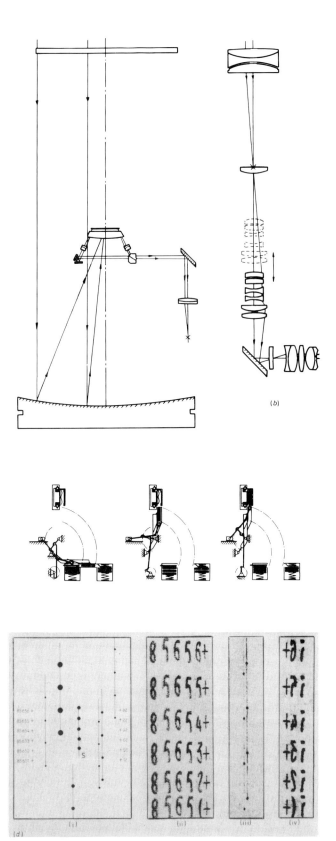

2.2.4

(*a*) V.E.B. Carl Zeiss Jena automatic camera for astrogeodesy. The Schmidt optics have an aperture of 500 mm and the mirror is 788 mm diameter. The eyepiece has a zoom lens for examination of the image on the 90 mm × 120 mm photographic plate. On the observatory wall is illustrated the Crab Nebula expanding cloud of gas which was the site of a supernova explosion observed in A.D. 1054 by Chinese astronomers. (*b*) Schematic diagram of Schmidt camera optics. (*c*) Principle by which the photographic plates are automatically changed in the Schmidt camera, and illustration of the shutter which can be opened and closed by remote control. The three positions show how a plate is taken from the magazine and positioned in the image plane. The exchange of an exposed plate from the cassette for a new plate takes about 15 s. (*d*) Illustration (i) shows an exposed camera plate with S identifying satellite-measuring points. The other picture points are of 'fixed' stars. The numbers in the plate margins are map references. Illustrations (ii), (iii) and (iv) are microphotographs of the actual records received by a Schmidt camera.

Grubb–Parsons

Sir Howard Grubb, Parsons and Co. Ltd, Newcastle upon Tyne, have had a long experience in the manufacture of large astronomical telescopes (see section 1.5). In 1976, under a contract from the Science Research Council, a 150 in (3·8 m) infrared mirror was ground and polished for a telescope to be installed in an observatory on the summit of Mauna Kea (4200 m) on the island of Hawaii (Figs 2.2.5 and 2.2.6).

With the permission of Mr G. J. Carpenter of the Royal Observatory, Edinburgh, some details are given of this new flux-collecting telescope designed specifically for infrared wavelengths. The 3·8 m flux collector was in-tended to be a low-cost special-purpose design with 2 seconds of arc on-axis image quality.

The paraboloid–hyperboloid Cassegrain mirror system includes an $f/9$ conventional Cassegrain, an $f/35$ chopping Cassegrain, an $f/20$ Coudé and an $f/2·5$ folded prime mirror. The primary mirror is made of very thin Cervit material, only 295 mm at the edge, and so there are engineering problems in the mounting.

The radial support system for the primary mirror is a conventional weighted lever-arm system, but the axial support is from 80 pneumatic cylinders disposed in three rings. These cylinders are all at the same pressure and controlled by three load cells on the neutral axis on the underside of the mirror.

2.2.5
(a) Thin, low-weight (13 500 lb), 150 in (3·8 m), primary mirror for infrared telescope being polished on Grubb–Parsons 240 in (6 m) polishing machine at Newcastle upon Tyne.
(b) Computer-controlled polishing machine for astronomical telescope mirrors. (c) Figuring of 150 in (3·8 m) mirror objective.

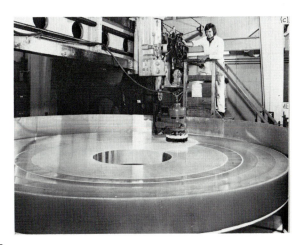

The central 1 m hole was etched with hydrofluoric acid to eliminate fine scratches which would form stress points during the centre-lift handling of the mirror at the time of installation.

An English type of yoke mount is used, appropriate to a latitude of some 20°, but possible problems may arise because of the difficulty of making azimuth north–south adjustments and differential movements which may be caused by earthquake conditions in Hawaii.

A computer system provides the means for pointing the telescope within a setting accuracy better than 30 seconds of arc and an overall guiding accuracy of 2 seconds of arc. The Royal Observatory, Edinburgh, was responsible for the construction of this telescope and is also responsible for observations after it was commissioned in 1978.

Infrared astronomy includes the electromagnetic radiation spectrum from 750 nm to 3 mm wavelength, which is the beginning of microwave radio astronomy. Much of the infrared will remain unexplored from Earth-based instruments because of absorption by the Earth's atmosphere. The region from 750 nm to 1·2 μm wavelength is named the photographic infrared because emulsions will respond to this radiation.

All observations beyond 1·2 μm wavelength are currently made with single point image detectors. The region from 1·2 to 5·2 μm is known as the near-infrared. From 5·2 to 8 μm the Earth's electromagnetic waves are absorbed.

There is a window between 8 and 14 μm and another from 17 to 22 μm wavelength of interest for astronomical measurements. The region from 22 to 1000 μm wavelength, which is mostly opaque, may have windows but the 1 to 3 mm window will close the gap between infrared astronomy and short-wave radio astronomy.

It has been estimated that there are about 20 000 infrared sources, most of which are stars, and about 5500 are at least 2·5 times the minimal detectable brightness. This compares with about 6000 stars which can be counted in a visual survey with the unaided eye at the same latitude.

Because the peak of our sun's energy falls near the middle of the 400 to 700 nm wavelength visible region, any stars which are cooler than the Sun will emit most of their energy as infrared beyond 700 nm wavelength. Stars as cool as 3000 K are well known and others have been found between 1000 and 2000 K. The coolest stars—red dwarfs and the

2.2.6
Photograph of early model of the 150 in infrared telescope for installation at Mauna Kea Observatory, Hawaii, during 1978.

dark companions of hotter stars—account for most of the stellar matter in space (Fig. 2.2.7).

The atmosphere of the Earth emits infrared radiation, which is particularly strong in the 8 to 14 μm region, so astronomers must take into account this background brightness which can be compared to daylight for visible-light astronomers who may be trying to observe stars by day. A cooled and shielded thermal detector was developed in 1960 which operates at a few degrees Kelvin. The bolometer is situated at the bottom of a cylindrical metal shield, which is refrigerated by liquid nitrogen, so few infrared photons can reach the detector except through the field of view of the cylinder.

A semiconductor crystal detector, known as a quantum infrared detector, has more recently been developed, which will absorb individual photons and release a unit of electrical charge for each photon, so conductivity can be used to measure infrared flux. The detector is exposed alternately to two beams. One beam consists of infrared from the star being observed and the surrounding sky, whilst the other beam is only from the sky. A rotating chopper blade determines which beam reaches the detector and, if the flux from the two beams is different, the detector will produce a signal which is proportional to the difference in energy. No visual observations are made with the telescope and all detection is by electronic equipment.

A rectified signal provides a direct-current voltage which can be recorded and, after calibration, can be used to compute the infrared flux received by the aperture of the telescope.

2.2.7
Blackbody radiation. The curves show the distribution of radiance with wavelength and satisfy Planck's law:

$$N_\lambda = A\lambda^{-5}\{\exp[(B/\lambda T)-1]\}^{-1}.$$

The straight line joining the maxima illustrates Wien's law:

$$T\lambda_{N_\lambda \max} = 2898 \ \mu m \ K.$$

The total emission, related to the integral of the expressions for the curves (i.e. the area beneath them), can be found from the Stefan–Boltzmann law:

$$W = \sigma T^4.$$

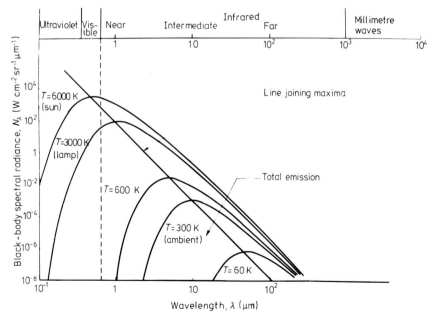

61

2.3 Night telescopes, image intensifiers and fibre optics

An image intensifier is an instrument that will produce observable output images that are brighter than the low-level input images. Optical image intensifiers used in passive systems rely for their operation on available light which may be at an extremely low level, such as that from a star or moonlight (Table 2.3.1).

Table 2.3.1

Natural illumination	Source intensity (lm ft^{-2})
Full sunlight	1000 to 100 000
Overcast day	10 to 1000
Twilight	1
Deep twilight	0·1
Full moonlight (clear)	0·01 to 0·1
Full moonlight (overcast)	0·001 to 0·01
Quarter moon (clear)	0·001
Starlight (clear)	0·0001
Starlight (overcast)	0·00001

The output from an image intensifier may be viewed directly through an eyepiece or indirectly by a camera and photography. Second-generation large-aperture image intensifier tubes, magnetically focused, will amplify the incident image to a level necessary for the correct exposure of daylight film emulsions in a camera with the response extended from visible into the near-infrared at 700 to 900 nm wavelength.

Supplementary illumination may sometimes be used in the form of an infrared torch or flashlight to improve vision under very low light conditions or to avoid excessive image-degrading photographic exposure times. Lenses should be optimised for the near-infrared with a reasonable performance in visible light. Optical design must include the faceplate of the image tube, which consists of glass fibres fused into a solid plate, and the photocathode coating.

Fibre optics were first reported in *Nature* during 1954 by A. G. S. van Heel and also by H. H. Hopkins and N. S. Kapany as a new method of transmitting optical images. The earliest attempts to manufacture glass fibres were made by the Egyptians about 3500 years ago and very small-diameter glass and quartz fibres were made by Michael Faraday in 1832. John Logie Baird patented a device incorporating the principles of fibre optics in 1926, but it was 30 years later before fibre optics became a practicable commercial development.

Optical fibres consist of a core of glass having a high refractive index, surrounded by a sheath of glass with a lower index. A beam of light incident on one end passes down the core by a series of total internal reflections at the interface with the sheath into which it penetrates about half a wavelength during reflections. Any scratches in the sheath which come within half a wavelength of the core surface will allow light to escape from the fibre. Further information is published in *Optical production technology* (Fig. 2.3.1).

2.3.1
Fibre optics is a term which loosely describes a group of products that are capable of transmitting light around corners. The products are made up of large numbers of very small fibres.

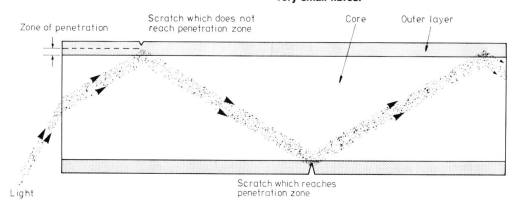

Zone of penetration — Scratch which does not reach penetration zone — Core — Outer layer — Light — Scratch which reaches penetration zone

2.3.2
A non-coherent bundle—also known as an incoherent bundle—light guide or light pipe consists of clad fibres randomly grouped together. In a coherent bundle (or image conduit), the relative positions of the fibres are the same at both ends. In consequence, an image formed on one end by a lens system is transmitted unchanged to the other, each fibre carrying one element of the picture.

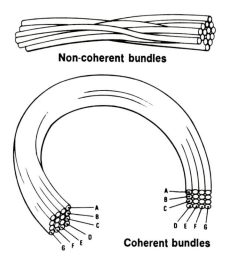

Non-coherent bundles

Coherent bundles

2.3.3
Non-coherent flexible fibre optic components where the fibres are aligned at the ends only and free along the length.

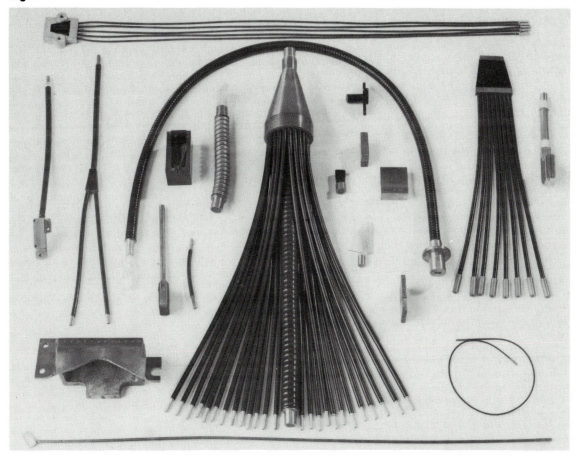

63

Glass-coated glass-core fibres are commonly used for light guides and fibrescopes, when the core refractive index is 1·62 and the sheath coating has refractive index 1·52 giving a numerical aperture of 0·56 (Figs 2.3.2 and 2.3.3).

Fibre optic faceplates for image intensifiers have a high optical efficiency and special forms can be produced by twisting, stretching or tapering the fused fibre bundles. The core refractive index is usually 1·81 with a sheath coating refractive index of 1·48 giving a numerical aperture of 1·04. Faceplates are a rigid form of coherent fibre bundle in which a large number of precisely aligned elements are fused together to give the required high image resolution. They must be vacuum tight and thermally compatible with the tube envelope materials as well as being chemically compatible with the phosphors or photocathodes which are to be deposited on them. Fibre sizes range down to a few micrometres (μm) diameter in high-resolution faceplates and so very large numbers of fibres are required to make up a bundle (Figs 2.3.4 and 2.3.5).

Night telescopes consist of an objective lens, which forms an optical image on the input surface of a fibre optic faceplate, an intensifier tube and an eyepiece. The faceplate fibres transmit light to a photocathode, on the rear side of the faceplate, where photons are converted into electrons which are accelerated and focused by an electric field to impact on a phosphor screen with greatly increased energy.

In the first-generation system, a fluorescent phosphor screen was on the inner surface of a fibre optic output window, and the impacting electrons were converted to visible light which passed through the fibres and was viewed through an eyepiece. A gain or light amplification of about 50 was achieved with an accelerating voltage of 15 kV.

A larger gain was possible by cascading several stages, and if three stages were coupled together then 50 × 50 × 50 or 125 000 times gain was theoretically possible but, due to coupling losses, there was considerable image bloom and persistence which limited the use of this system in the presence of bright light sources.

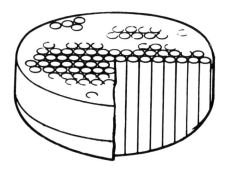

2.3.4
Fibre optic faceplates having plano–convex or plano–concave surfaces can be used to flatten an image on a curved surface. A faceplate is made from a large number of short fibres aligned side by side and fused together to form a solid plate. They can be used as faceplates for cathode ray tubes and image intensifiers.

Mullard Ltd has pioneered the development of cascade image intensifiers and in 1968 publicised the second-generation microchannel image intensifier which does not have these coupling disadvantages (Fig. 2.3.6).

The second-generation image intensifiers are similar to the first-generation ones but with a microchannel plate electron multiplier placed ahead of the phosphor viewing screen (Fig. 2.3.7).

The microchannel electron multiplier will give electron gains of several thousands depending on the voltage across the plate and, because the plate is close to the phosphor screen, electrons arrive without any significant spreading. The image is inverted so the intensifier can be used directly with a lens.

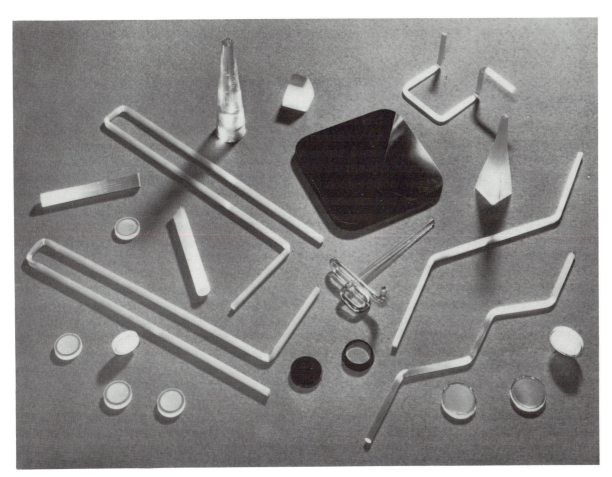

2.3.5
Coherent solid fibre optics where the bundles are fused
together to form rigid components such as image
inverters, magnifiers, minifiers, faceplates and field
flatteners.

2.3.6
Second-generation microchannel image intensifier.

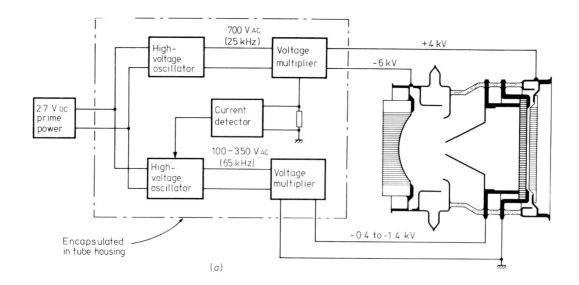

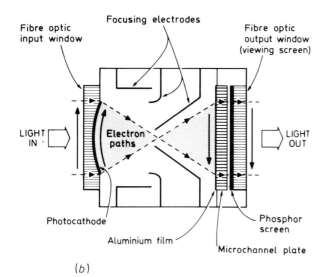

2.3.7
Electrostatically focused microchannel inverter image intensifier. (a) Circuit diagram—the only connection needed is to a single 2·7 V battery. (b) Detail of the image intensifier. (c) A channel plate image intensifier uses an assembly of small tubes made with semiconducting glass. Each channel, by virtue of a high potential applied between its ends and an emissive internal coating, acts as an electron multiplier. The density of the electron image on the output side of the plate is therefore much greater than on the input side, and a bright image is formed on the luminescent viewing screen.

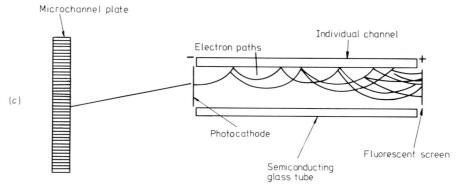

66

3 Microscopes

3.1 Introduction

The invention of the microscope is generally attributed to Zacharias Janssen in about 1590 and to Galileo Galilei (1564–1642) in 1610. It is uncertain as to exactly who first invented the microscope, but the Dutch microscopist Anton van Leeuwenhoek (1632–1723) made biconvex lenses, of very small diameter and short focal length, which enabled him to carry out scientific investigations and was the first to give publicity to microscopy. Robert Hooke (1635–1703) published his work in *Micrographia* during 1665 which described a compound microscope and observations made with it (Fig. 3.1.1).

Between 1650 and 1800, the mechanical parts of the microscope were developed, but the only optical improvement was made by William Hyde Wollaston (1766–1822) who combined two plano–convex lenses, in the form of a doublet with an air gap between them, in which he placed a diaphragm which balanced spherical aberrations and gave a relatively large field of view.

The technology of making achromatic microscope objectives was more difficult than producing achromatic telescope lenses. The first achromatic microscope objectives were manufactured in 1807 by Hermann van Deyl (1738–1809) and his lens had a focal length of 13 mm with a magnification of 19×. The elimination of chromatic aberrations did not remove spherical aberration, which was troublesome when powerful eyepieces were used.

Giovanni Battista Amici (1786–1863) had been making achromatic lenses since 1816 and, following up some suggestions made by Joseph Jackson Lister (1786–1869) in 1830, he introduced the hemispherical front lens. With his objectives, he was able to make the aperture angle in air as high as 120° and drew attention to the effect of the coverglass thickness on

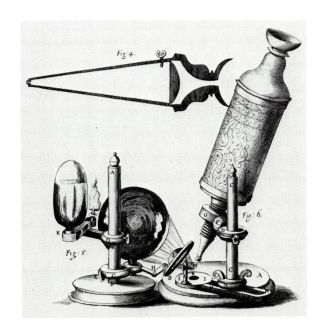

3.1.1
The Hooke microscope (1665).

spherical aberration. Amici introduced cedar oil immersion in 1850 which had the advantages of increasing the numerical aperture, reducing the light loss, and eliminating spherical aberration and the effect of coverglass thickness.

To reduce colour in the image, Ernst Abbe (1840–1905) tried to eliminate the remaining longitudinal chromatic aberration as well as the differences in spherical aberration. By cooperation with Otto Schott (1851–1935), a new optical glass was developed in Jena which enabled the apochromat objective to be produced in 1886 and, with the use of special compensating eyepieces, chromatic aberration was eliminated.

The compound microscope uses two stages of magnification. The total magnification is the product of the objective, which is near the specimen, and the eyepiece or ocular, which amplifies the intermediate image formed by the objective. The eyepiece is usually designed so that there are aberrations in the lens which will balance uncorrected aberrations in the objective (Fig. 3.1.2).

The eyepiece must provide a virtual image of the intermediate image, at about 250 mm distance, so that it can be viewed in comfort by the observer with a relaxed eye. The centre of the exit pupil should be at some convenient distance at least 10 mm from the last lens surface.

The Huygens eyepiece, named after Christiaan Huygens (1629–95), is a low-cost system which consists of a field lens positioned before the virtual image, formed by the objective, and brings it into focus within the eyepiece combination. A field stop is used to control the aperture and form the support for a graticule scale.

The focal plane of the eye lens, where the real image is formed, is situated between the two lenses at the field stop. The eye relief of this type of lens is only about 3 mm, so the observer cannot use spectacles, and although focal lengths can be adjusted by movement of the eyepieces there can be no compensation for astigmatism of the eye (Fig. 3.1.3).

The Ramsden eyepiece, designed by Jesse Ramsden (1735–1800), has a principal focus in

3.1.2

Optical and mechanical features of a compound microscope by the American Optical Corporation, Buffalo, N.Y., showing Köhler-type illumination and wide-field eyepiece.

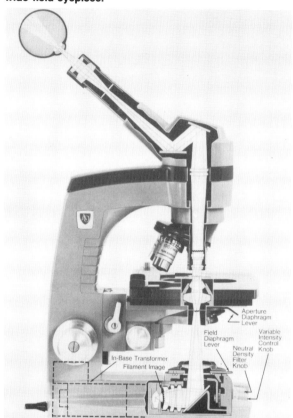

front of the field lens and has the advantage that graticules, cross wires, grids or micrometers can be mounted on the eyepiece holder outside the field lens. As the graticule and intermediate image are in the same plane, they will both be in focus. For these reasons, and because the eye relief is about 12 mm, the Ramsden eyepiece is relatively popular, although the field of view is usually less and the achromatism not as good as that of a Huygens eyepiece of the same magnification (Fig. 3.1.4).

The Kellner eyepiece, designed by Dr H. Kellner in 1849, is essentially an achromatised Ramsden and is used in moderately wide-field instruments, as it provides a wider and flatter field and also has greater eye relief to give more comfort to wearers of spectacles (Fig. 3.1.5).

68

3.1.3
The Huygens eyepiece.

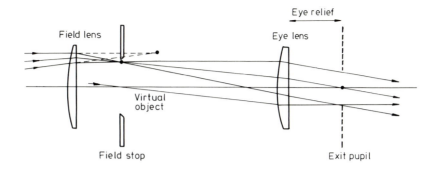

3.1.4
The Ramsden eyepiece.

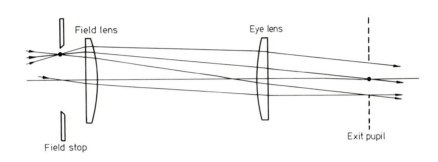

3.1.5
The Kellner eyepiece.

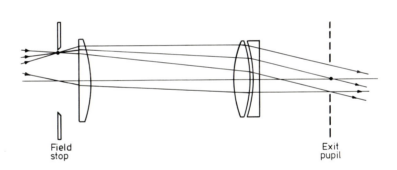

3.1.6
The orthoscopic eyepiece.

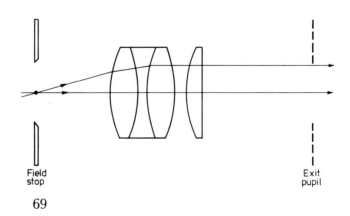

69

The orthoscopic eyepiece, which is corrected for distortion, has a wide field, high magnification and eye relief of about 20 mm, so it is often used on high-quality instruments where low cost is not of prime importance (Fig. 3.1.6).

An inclined binocular body has parallel eyetubes which permit full relaxation for the eye muscles used in binocular vision. One eyetube has a focusing sleeve in order to provide a separate correction if the observer's eyes are different from each other. A tube length correction lens compensates for the additional optical path length and maintains the correct conditions for which standard objectives are computed.

Some microscopes have zoom lenses. By varying magnification and moving the image plane, the system can be adjusted for optimum viewing within a magnification of about 5× factored by the magnification of the rest of the system.

Microscopy needs photography in order to provide communication between workers in a particular field, and the invention of the negative–positive photographic process, by William Henry Fox Talbot in 1835, was of great importance. Before this invention, microscopists could only compare observations by means of drawings or by sending the actual specimens by letter post.

Stereomicroscopes are designed as two distinct types. The twin-objective Greenough design, first manufactured by Zeiss in 1897, has two completely separate beam paths. This type can have compact pairs of objectives with slender mounts, so that the specimen is readily accessible, and the chromatic aberrations can be easily corrected. The converging tubes provide fields of view that are inclined at about 15° to each other, and Porro prisms are used to erect the images which appear to be about 250 mm from the eyes (Fig. 3.1.7).

A disadvantage of this design is that the two intermediate image planes are tilted relative to one another, so only the median bands are sharply focused at the same time. This can be a cause of operator fatigue. Photographs can be taken only when the specimen is tilted or a beam path is tilted relative to the specimen.

3.1.7
Schematic beam path of a Greenough stereomicroscope.

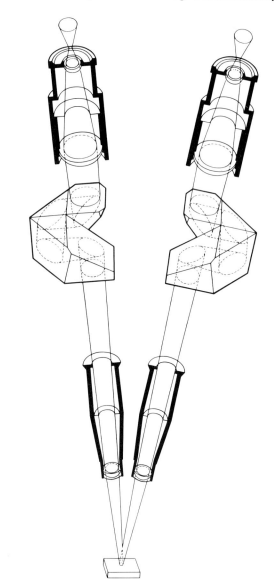

The alternative design of stereomicroscope has a common main objective and the intermediate image planes are parallel to the objective plane. The stereoscopic effect can be easily varied and zoom magnifications can be changed without altering the working distance or focusing the objectives. A surface perpendicular to the axis of the instrument is perpendicular to the optical axis in both tubes, so both fields of view can be imaged sharply and observations are less tiring. Photomicrography or projection accessories can be easily fitted to this type of stereomicroscope. A disadvantage is that the two beam paths traverse the common main objective obliquely, so chromatic aberrations are difficult to correct; also, the lenses become relatively large and expensive (Fig. 3.1.8).

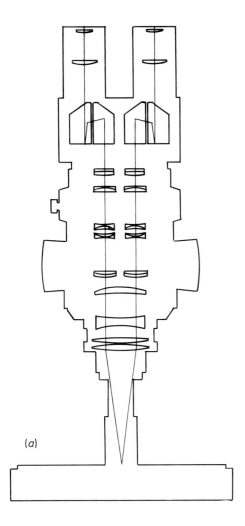

(a)

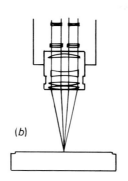

(b)

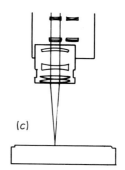

(c)

3.1.8
(a) Optical design of the Wild M7A Stereomicroscope with zoom magnification changer. (b) Wild M7S: main objective positioned for stereoscopic observation. (c) Wild M7S: main objective positioned for vertical observation. The possibility of vertical observation is particularly advantageous in photomicrography, because the vertical beam path enables the well corrected central part of the objective to be used to produce a flat, high-quality image.

3.2 Microscope optics and illuminators

When an object is sighted with the eyes, an image is formed on the retina, and the dimensions and distance of an object determine the size of the image. The visual angle at the retina is usually the geometrical angle of vision. The physiological limiting angle in good illumination is about 1 minute of arc, at 250 mm visual distance, and details at smaller angles cannot be detected without magnification by a microscope which enables a very small object to be seen by the eye under an extended visual angle (Fig. 3.2.1).

Objective lenses

At the first stage of image formation, an objective lens produces a magnified real image whose quality affects the general performance of the instrument. The inverted real aerial

71

image is called an intermediate image, and can be seen without an eyepiece on a ground-glass screen. A second stage consists of viewing the intermediate image through an eyepiece, which acts as a magnifier so that a virtual image of the object can be seen by the eye (Fig. 3.2.2).

A compound microscope has a magnification which is the product of the objective and eyepiece magnifications. A set of matched objectives is required for the total range of a microscope and, as these are parfocal, when they are mounted on a revolving nosepiece the focus is retained after changing objectives or eyepieces.

The ability of an objective lens to resolve fine detail depends on its magnification ratio

3.2.2
Optical path in the microscope. The objective forms a magnified, real, inverted and side-reversed image of object P at the reproduction scale 5:1. The subsequent 8× eyepiece magnifies this intermediate image another 8 times. The observer thus sees the image as if he viewed the 40 times (i.e. 5 × 8) magnified object from a distance of 250 mm without an instrument. The magnification in the diagram is not to scale.

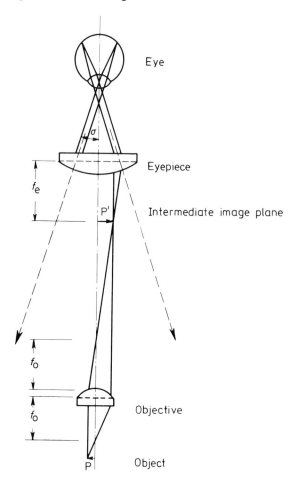

3.2.1
The black arrows represent an object situated at various distances from the eye. The angles between the broken lines and the horizontal axis are the angles of vision when no lens is used for observation. The angle of vision increases if the object is brought nearer to the eye, and the image on the retina of the eye also increases in size. If a lens is placed in front of the eye and the object is at the focal point F (a distance f from the lens, f being the focal length), the object will be imaged at infinity. The lens has thus increased the angle of vision. This results in a magnified image on the retina of the eye. The magnification M is given by $M = 250/f$, i.e. the object appears to be linearly enlarged by this amount compared with its apparent size when viewed by the unaided eye from the conventional viewing distance of 250 mm. In the figure, f is the focal length (in mm) on the object side of the lens, f' is the focal length (in mm) on the image side of the lens, H is the principal plane of the lens, F is the focal point on the object side, and F' is the focal point on the image side.

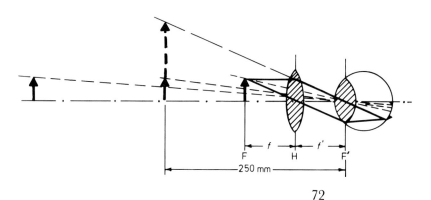

and numerical aperture, defined as the product of the refractive index of the optical medium and the sine of the angle included between the marginal ray accepted by the objective and the optical axis. High-power objectives usually have the largest numerical apertures and this determines their ability to resolve closely spaced lines or points. The useful total magnification of a microscope is in the range of 500× to 1000× the numerical aperture of the objective (Fig. 3.2.3).

The resolving power is determined entirely by the design of the objective, but the size of the field of view depends on the characteristics of the eyepiece and on how much of the intermediate image is covered. A Huygens eyepiece may have a 30° angle of view, whereas a 50° angle may be achieved with a wide-field eyepiece (Figs 3.1.3 and 3.1.6).

Axial chromatic aberrations must be corrected whenever white light is used, or any light other than monochromatic, as a lens disperses the rays into various colours, refracting

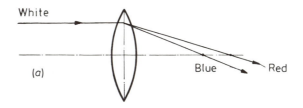

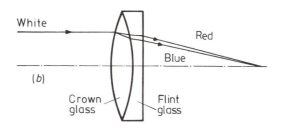

3.2.4
(a) Chromatic aberration. (b) Correction of chromatic aberration.

3.2.3
The numerical aperture of the objective equals _n_ sin _u_, where _n_ represents the refractive index of the medium betweeen the coverglass and the front lens, and _u_ is half the angle of aperture of the objective. The numerical aperture is a measure of the amount of light accepted by the objective. When the medium between the front lens and the coverglass is air, then _n_ = 1. In this case, therefore, the numerical aperture _A_ of an objective must always be less than 1, because for practical reasons the half-angle of aperture _u_ for air can never be as much as 90°. Objectives designed for this technique are called dry systems. Therefore, the numerical aperture of a dry system is always less than 1 and in practice does not exceed 0·95.

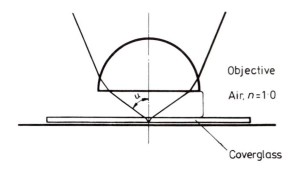

red light least and violet most of all. Red 656 nm, green 546 nm, blue 486 nm and violet 405 nm wavelength spectral lines are used during colour-correction calculations. The combination of flint and crown glasses makes it possible to match lenses so that the focal length of green light coincides with that of red light and enables us to make a normal achromat. If we must also match the focal length of blue light then we can use fluorspar for one lens as it refracts more strongly than glass.

Fluorite is a form of calcium fluoride which is called fluorspar in its transparent form. It is crystalline in structure, but acts as if it were isotropic, and is not birefringent. Fluorspar is easily worked and transmits to nearly 100 nm wavelength, so is suitable for ultraviolet optical elements and can be used to achromatise quartz. Fluorspar is suitable for lenses in semi-apochromats and will reduce the secondary spectrum in microscope objectives (Fig. 3.2.4)—see _Optical production technology_ and _Spectacle lens technology_ for further details.

Lateral chromatic aberrations show themselves as variations in magnification of the different colours and these increase with distance of an object point from the optical axis. If axial chromatic aberrations are eliminated but lateral chromatic aberrations are uncorrected, the eye would see an image in pure colours surrounded by weak colour fringes. Lateral chromatic aberration is eliminated by the use of cemented lenses in an eyepiece and also by computing the same chromatic difference of magnification for all objectives in a range. Eyepieces are usually designed to compensate for the chromatic difference in magnification produced in a range of objective lenses.

Spherical aberration increases with the aperture of a lens because rays of light passing through a lens have different intercept lengths depending on their level of passage compared to the optical axis. Light is refracted more strongly across the margin portions than near the centre, so the rays do not intersect at a point. Spherical aberration cannot be completely eliminated from spherical surfaces, but can be corrected by a combination of flint and crown glasses used for collecting and dispersing the rays of light (Fig. 3.2.5).

Astigmatism is an aberration which increases with distance of an object point from the optical axis, so that two image fields of different curvature are produced by a plane object. If an objective is not corrected for astigmatism, there will be two image points at different distances from the image plane and so the image will not be sharp in the marginal regions (Fig. 3.2.6).

Correction for astigmatism can be achieved by making the two image fields with an identical curvature, but this field curvature cannot be eliminated for microscope objectives of conventional design. Residual field curvature can

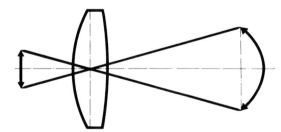

3.2.6
Astigmatism.

be compensated by the use of a negative lens and green-yellow filter for black and white photomicrography, but this method is unsuitable for visual observations due to the introduction of chromatic aberrations (Fig. 3.2.7).

Coma and distortion are two further aberrations that become noticeable towards the marginal regions of the image field. Coma is the asymmetrical spherical aberration of the oblique ray bundles which take the form of image dots with tails resembling comets. Distortion is identified by the reproduction ratio not being constant over the entire image. An image in the form of a square is reproduced as the shape of a cushion or barrel depending on the type of lens.

When coverglasses are used, there will be an overcorrection of spherical aberration and this is noticeable with objectives above 0·4 numerical aperture. Spherical aberrations must be compensated by undercorrection of the objective lens and, because the coverglass becomes a component of the image-forming system, the thickness of the coverglass must be constant. A value of 0·17–0·18 mm has been adopted as a standard for microscopes when using transmitted light.

3.2.5
Spherical aberration.

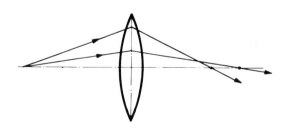

3.2.7
Field curvature.

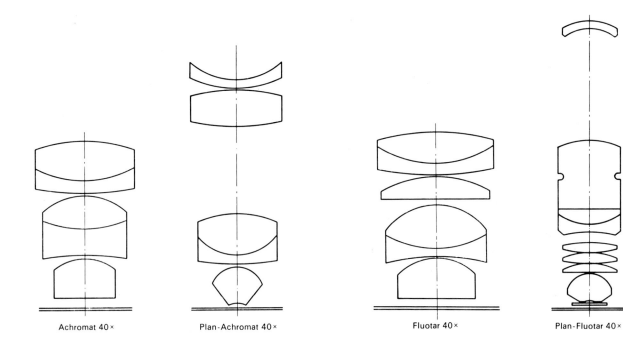

Achromat 40×   Plan-Achromat 40×   Fluotar 40×   Plan-Fluotar 40×

3.2.8
Six optical designs of microscope objectives.

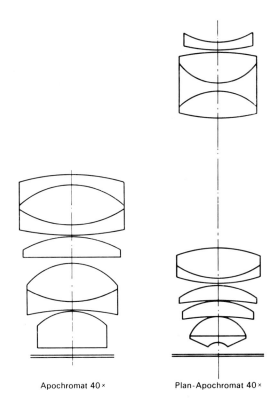

Apochromat 40×   Plan-Apochromat 40×

Immersion objectives are not so sensitive to coverglass deviations, as they are compensated by a variation in the oil film thickness. Objectives for use with incident light are usually computed for use without a coverglass.

Microscope objectives are classified according to their state of correction and, as colour correction is improved, the lenses become more complicated in design.

Achromats are chromatically and spherically corrected objectives in which the chromatic aberration has been fully removed for two colours, which are the red Fraunhofer C line of 656 nm wavelength and the green-blue Fraunhofer F line of 486 nm wavelength, within the maximum colour sensitivity of the human eye. For black and white photomicrography, a green-yellow filter is advantageous to absorb violet radiation to which the photographic material is sensitive.

Semi-apochromats are objectives which use fluorspar lenses, to extend the spectral range, and provide a larger aperture than corresponding achromats which use only crown and flint glass lenses. Fluorite systems are suitable for fine-detail research work and colour photomicrography.

Apochromats have chromatic aberrations corrected for three spectral colours to provide the ultimate performance and elimination of residual chromatic aberration for research investigations.

Plano objectives are designed for flat-field photomicrography and projection where a curved field, as provided by conventional microscope objectives, will not be satisfactory. Suitably shaped negative lenses of converging effect have made it possible to flatten the field of view. Unfortunately, the means of flattening the field counteracts the apochromatic correction of the objective and it is necessary to use fluorspar for the largest possible number of elements (Fig. 3.2.8).

Condenser systems

The illumination system has a decisive influence on the sharpness and general character of microscope images. A light source must deliver rays, as required by the objective and eyepiece in use. Also, the field of view and exit pupil of the objective should be uniformly illuminated. Facilities must provide for changing the cross section of the rays in the rear focal plane of the objective.

The relatively large fields of low-power objectives must be uniformly illuminated, but the condenser system should also provide for adapting its aperture to that of the objective.

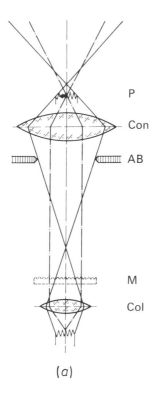

(a)

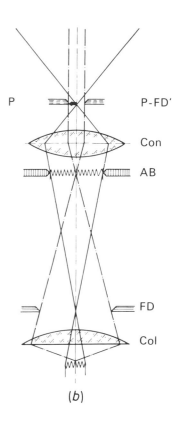

(b)

3.2.9

(a) Critical illumination. In critical illumination, the light rays emerging from the filament become parallel once they have passed through the collector lens (Col) and are then concentrated on the specimen (P) by the condenser (Con). The filament thereby becomes sharply imaged in the specimen plane. Its structure is visible, although the field of view is unevenly illuminated. The image of the filament can be eliminated by placing a frosted or opal glass filter (M) in front of the collector lens. The background of the field of view then becomes evenly

illuminated. *(b) Köhler illumination.* The filament is imaged in the condenser diaphragm plane AB by the use of the collector lens (Col). Normally, however, it suffices to project an image of the filament onto the microscope mirror. The field diaphragm (FD) of the lamp is imaged in the specimen plane (P–FD') by means of the condenser (Con), and is then opened until the entire field of view is illuminated. If the field diaphragm is opened too far, the resulting excess light will cause the structures to become unsharp, and some contrast will also be lost.

These variations are achieved by means of a condenser iris diaphragm in the illuminating beam, the image of the diaphragm thus limiting the rays in the objective. The limiting aperture should be as large as that of the highest-power objective. The Abbe two-lens critical condenser is not corrected for spherical or chromatic aberration but is suitable for general observation (Fig. 3.2.9).

A second diaphragm, known as the lamp field diaphragm, limits the ray cross section to the required diameter and should match the exit pupil. Köhler (1893) illumination has a source which is imaged in the aperture of the substage condenser. The first iris diaphragm controls the area of the object to be illuminated, and the second diaphragm varies the numerical aperture of the illuminating beam. The field diaphragm should be used with bright-field, dark-ground or phase-contrast illumination and is of importance for colour photomicrography.

In dark-ground microscopy, the image is formed by light diffracted on the object. Undiffracted light does not enter the objective so blank areas of a microscope slide appear dark in the field of view, and objects appear bright against a dark background.

3.3 Biological and metallurgical microscopes

The American Optical Series 10 'Microstar' advanced laboratory microscope has an infinity-corrected system and Köhler-type illumination. The infinity-corrected system makes it possible to change the objective-to-eyepiece distance without affecting optical performance, and this simplifies the focusing mechanism by adjustment of the nosepiece only and not the stage assembly or body tube (Fig. 3.1.2).

Immersion oil used in a biological microscope must have a suitable refractive index and dispersion, it should be chemically inert, free from a tendency to spread or creep, remain fluid and not harden rapidly when exposed to air, and its optical properties should be stable. Cedarwood oil, Crown oil or Shillaber's oil can be used with immersion objectives.

Wide-angle eyepieces contain a graticule, with the etched side facing outwards, and because the projected values of the graduations vary with the optical combination, they should be calibrated before accurate measurements are made.

Calibration is achieved by focusing on a stage micrometer and then moving it until one of the graduations corresponds exactly with one of the divisions of the eyepiece micrometer. The true distance (X) seen on the stage micrometer, which corresponds to the number of divisions (Y) of the eyepiece micrometer disc, is then read, and by dividing this true distance by the number of divisions of the eyepiece micrometer it is possible to find the distance that each division subtends (Fig. 3.3.1).

The Vickers M17 wide-field microscope can be equipped for dark-ground, phase-contrast and simple polarising applications. Köhler illumination is provided by a tungsten–halogen lamp, and where required a mercury-vapour lamp will supply ultraviolet or blue light. Cameras can be fitted for photomicrography (Fig. 3.3.2).

The Wild M12 microscope is specially designed for incident-light observations. Conventional microscopy, using transmitted-light illumination, is concerned with transparent slides of standard size and thickness. Industrial microscopy, however, often deals with non-transparent surfaces, both polished and unpolished. To be observed through the optical microscope, such surfaces require perpendicular illumination from above, in which case the objective lens also acts as a condensing lens. This system is termed bright-field incident-light illumination. Alternatively, oblique incident-light illumination may be required, whereby part of the light enters the lens after being reflected and scattered at various angles by the specimen surface If essentially no part of the reflected original main beam enters the objective lens, then true dark-field illumination conditions are attained and the visible detail

3.3.1

(a) Wide-field eyepiece. **(b)** Calibration of micrometer discs on reticle of wide-angle eyepiece for American Optical 'Microstar' microscope.

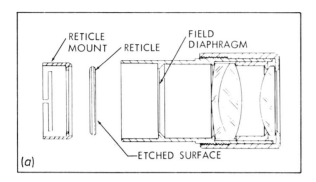

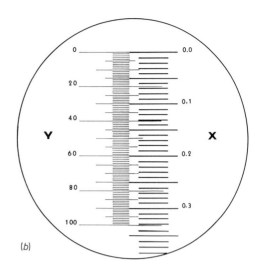

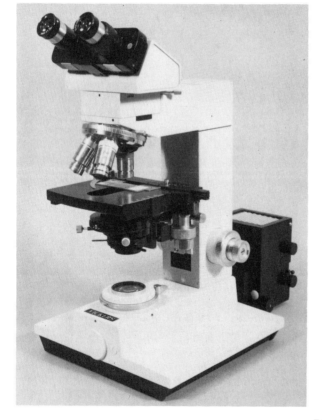

appears bright against a dark background (Fig. 3.3.3).

The Wild incident-light attachment enables the routine and research transmitted-light microscopes to be easily built up into full incident-light microscopes. The attachment is basically a drum-shaped device which fits readily between the microscope body and the monocular or binocular head. It incorporates all the necessary components for bright-field, dark-field and polarised-light observations. In addition, both conventional and incident-light microscopy can be carried out simultaneously under certain conditions for the inspection of semi-transparent objects (Fig. 3.3.4).

A special quadruple nosepiece is available for use with the incident-light attachment. The holes are larger than on the conventional nosepiece and will take the Universal Epi objectives, which are constructed for incident-light work in bright and dark fields. These five objectives ($4\times$, $10\times$, $20\times$, $40\times$ and $100\times$ oil immersion) incorporate ring condensers which

3.3.2
Vickers M17 wide-field biological microscope with transmitted illumination which can be easily modified for incident illumination to make it suitable for metallurgy and microelectronics.

78

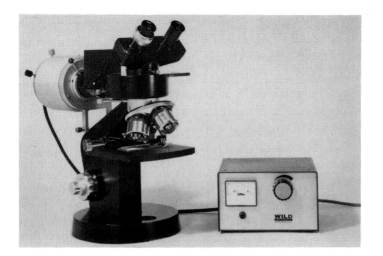

3.3.3
Wild M12 microscope with incident-light attachment and Universal Epi-achromat objectives.

ensure symmetrical illumination in incident-light dark-field conditions. If incident-light observations in bright field alone are required, the normal transmitted-light objectives up to 10× may be used in the conventional quadruple or sextuple nosepiece, as can also the 50×, 85×, and 100× oil immersion objectives. For medium- and high-power dry bright-field systems, however, special objectives corrected for use with uncovered specimens are required (Fig. 3.3.5).

The Wild M8 zoom stereomicroscope can be continuously adjusted within a zoom range of 1 : 8 and a total magnification of 6× up to 50×, when using a 10× eyepiece, which can be increased to the objective lens limits of 2·4× to 160× magnification with different power eyepieces and a supplementary objective (Fig. 3.3.6).

The angle of observation, or convergence subtended by a feature, is important but the viewing angle within which an eye can detect two separate features is critical. Details are resolved if they subtend an angle of between 2 and 4 minutes of arc at each eye, which represents a grid having between 7 and $3\frac{1}{2}$ lines per millimetre. With a 6× magnification, the eye should resolve 30 lines per millimetre, and with 10× up to 50 lines per millimetre (Fig. 3.3.7).

Visibility or Vernier acuity goes further than resolution, as a black line on a white background can be seen under good lighting conditions down to a viewing angle of about 4 seconds of arc or less at the retina. At 250 mm distance, a black line against a white background must be at least 5 μm (0·0002 in) wide in order to be seen by an unaided eye.

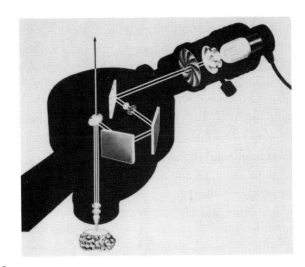

3.3.4
Light path in the Wild incident-light attachment at the bright-field setting.

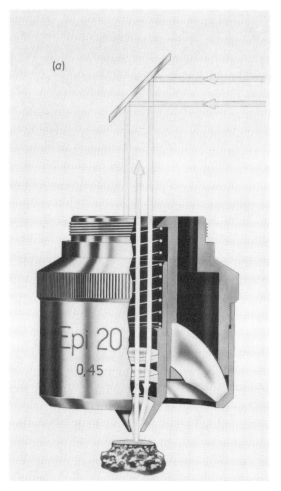

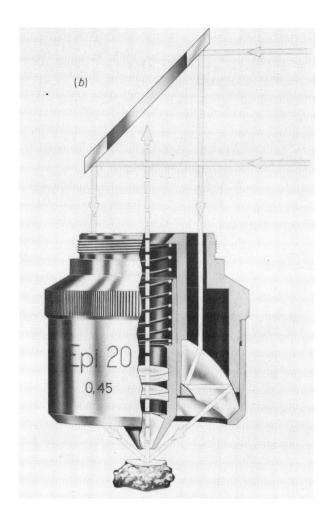

3.3.5
(*a*) Light path in the Universal Epi-achromat objective at the incident-light bright-field setting. (*b*) Light path in the Universal Epi-achromat objective at the incident-light dark-field setting.

At the conventional viewing distance, it is not the convergence angle itself that determines a perception of depth but the parallax angle between two points at different heights. The perception of depth at close range is very good, and a parallax angle of only 17 seconds of arc corresponds to a depth difference of 0·1 mm. In practice, unaided stereoscopic vision is limited to a range of 400 to 500 m, at which distance the convergence angle is about 2 minutes of arc. When the angle of parallax exceeds 70 minutes of arc, the impression of depth will disappear for the unaided eye.

Observation through a stereomicroscope is no more than an artificial increase in the normal viewing angle, so the ability to detect very small differences in depth is improved. Illumination and contrast relationships also affect the

80

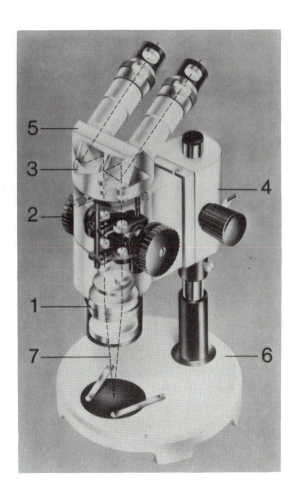

3.3.6
Wild M8 zoom stereomicroscope: 1, five-lens main objective; 2, zoom magnification changer 1:8; 3, image-erecting prism; 4, drive housing; 5, interchangeable binocular tube; 6, stand; 7, two separate optical trains are used to produce a spatial image of a three-dimensional specimen and provide an artificial increase over the normal viewing angle.

3.3.7
Binocular field of view. The angle of observation or convergence subtended by a feature being viewed at the conventional distance of 250 mm varies from about 12 to 17° for corresponding interpupillary distances of from 53 to 75 mm (α = convergence angle, σ = parallax angle, γ = viewing angle).

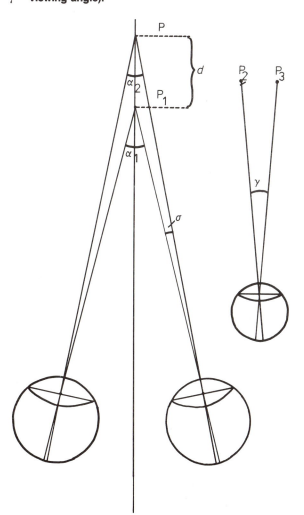

ability to see differences in depth or height of specimens under examination.

The advantages of a stereomicroscope with a common main objective are, first, that a surface lying perpendicular to the axis of the instrument is reproduced perpendicular to the optical axes in both tubes. Secondly, there is no tilt between the two intermediate image planes and so both fields of view can be sharply imaged. Finally, accessories for photomicrography, drawing or projection can be easily fitted. The disadvantages are that the main objectives are large and expensive, and that chromatic aberrations are difficult to correct due to the oblique paths of the light beams.

81

3.4 Photomicrography

The relationship between the darkest and lightest areas of a specimen is a problem to be overcome with all photomicrography. If there is strong contrast, then there is little difficulty, but weakly or moderately contrasted specimens will require carefully chosen filters, emulsions, exposure times and developers to ensure optimum performance.

In commercial photography, there will usually be a contrast range of about 1 : 30, but with photomicrographic subjects this range will only occur among highly coloured bright-field specimens. Usually the subjects have very weak contrast (Fig. 3.4.1).

The correct exposure time must be determined, and one method is to carry out test exposures with Polaroid film, which can be rapidly developed, and the data recalculated for photographic material of a different speed. If Polaroid emulsion of 400 ASA (27 DIN) is correctly exposed at $\frac{1}{50}$ s, then a black-and-white film of 25 ASA (15 DIN) must be exposed 400/25 or 16 times longer, i.e. for about $\frac{1}{3}$ s.

The spectral sensitivity of a photographic material varies according to the wavelength of light, and is different to the spectral sensitivity of the eye. The relationship between the quantity of light and the resulting blackening of the film can be obtained by placing a grey wedge of continuously increasing optical density on a strip of the film which is then exposed and developed. The photographic density can be measured with a densitometer.

The logarithm B of the product of light intensity I and exposure time t is plotted on the X axis. The light intensity is derived from the known properties of the light source and the optical wedge. The photographic density D of the exposed film is plotted on the Y axis and is defined as the logarithm of the reciprocal of film transmittance T. This gradation curve is known as the characteristic curve of the photographic material (Fig. 3.4.2).

The normal gradation $\gamma = 0.7$, representing a 35° angle of slope. If $\gamma > 0.7$ and the angle of slope $\alpha > 35°$, then the emulsion is said to be hard, and relatively small changes in exposure time produce large differences in blackening. If $\gamma < 0.7$, the emulsion is said to be soft, and large differences in specimen brightnesss can be reproduced.

The size, shape, distribution and number of silver grains in a photographic emulsion limit the reproduction of small features, and the resolution of normal emulsions lies between 60 and 120 lines/mm. Low-sensitivity emulsion has a high resolution and highly sensitive

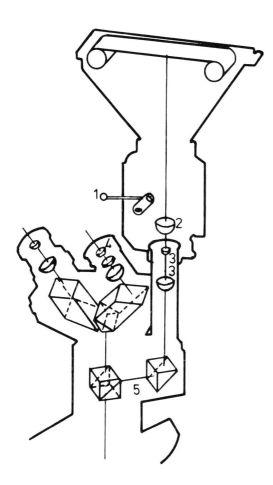

3.4.1

This attachable camera was designed specifically for use with Wild microscopes in trinocular assembly, and can be used for all formats from 35 mm to 4 in × 5 in. Beam path in trinocular photomicrographic assembly;
1, swing-in photocell; 2, compensating lens;
3, photo-eyepiece; 4, format-indicating eyepiece with graticule (reticle); 5, H phototube with beam-splitting prism.

82

3.4.2

Gradation curve of a film emulsion: I, fog; I–II, underexposure; II–III, blackening proportional to exposure (working range); III–IV, overexposure. The variables are: *I* **= intensity of illumination in the film plane (lx);** *t* **= exposure time (s);** α **= angle of slope;** $\gamma = \tan \alpha = \Delta D / \Delta B$**.**

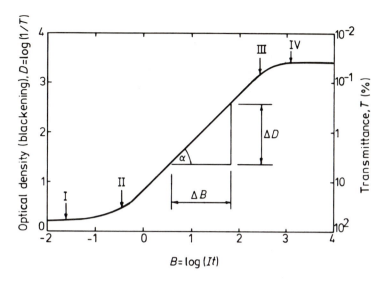

emulsion gives a low resolution.

Enlargement is limited by the size of the silver grains in the negative, and when the distance between the grains equals the resolving power of the eye, the limit has been reached. Grain size is influenced by the emulsion, by the exposure time (because overexposure produces a coarser grain) and by the development time. For best results, a low-sensitivity emulsion should be used, as this will give the highest resolution. Exposure time is of particular importance with colour emulsions as deviations of more than 50% from the correct exposure will produce incorrect colours.

3.5 Phase-contrast and interference microscopy

Phase contrast methods

Phase contrast is one of the most effective ways of observing unstained biological material under a microscope. In the phase-contrast microscope, light which is diffracted by a transparent specimen is made to interfere with direct undiffracted light to form an image with contrast. Staining techniques may alter or damage biological structures and cannot be applied to living material. Unstained poor-contrast specimens can be made visible by optical means alone using the phase-contrast techniques discovered by Professor F. Zernicke in 1942 (Fig. 3.5.1).

Stained specimens, which alter light amplitude, differ fundamentally from phase-altering specimens. The maxima of the diffraction images from an amplitude-altering specimen are of the same phase. Each structural element in the specimen absorbs light and the selected absorption produces a visible image (Fig. 3.5.2).

Unstained small living organisms are transparent and colourless. Because of the transparent nature of the fine detail within a phase specimen, the amplitudes of light waves passing through a specimen are almost unaltered, so a classical type of image which is visible to the eye cannot be formed.

Within the fine structural detail of the specimen, there are optical path differences that result from changes in refractive index. Some light is scattered, compared to the undeviated light which is transmitted directly where there is no detail to cause diffraction. A characteristic of diffraction is that there is a difference in phase between the diffracted and undeviated light, and this can be converted into

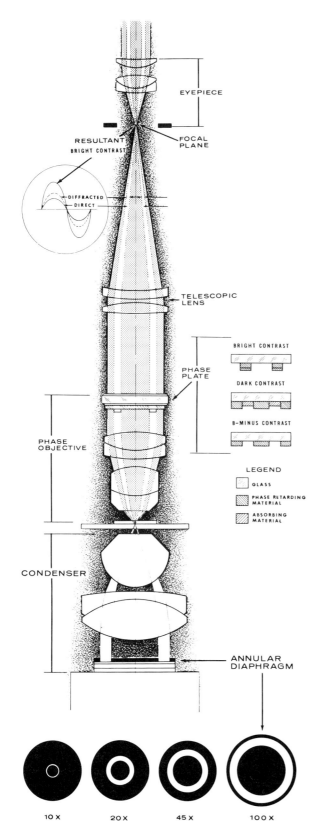

EYEPIECE

RESULTANT
BRIGHT CONTRAST

FOCAL
PLANE

DIFFRACTED

DIRECT

TELESCOPIC
LENS

PHASE
PLATE

BRIGHT CONTRAST

DARK CONTRAST

B-MINUS CONTRAST

PHASE
OBJECTIVE

LEGEND

GLASS

PHASE RETARDING
MATERIAL

ABSORBING
MATERIAL

CONDENSER

ANNULAR
DIAPHRAGM

10 X   20 X   45 X   100 X

3.5.1
Diagram of American Optical Series 10 phase microscope. Annular diaphragms below the condenser and a phase plate between the objective and eyepiece are features of phase microscopy.

amplitude differences which can be seen by the eye.

The phase microscope incorporates an annular diaphragm below the condenser which directs a hollow cone of light to the transparent specimen, and an objective which includes a diffraction or phase plate. This plate separates the diffracted and direct light coming from the specimen so that the intensity and phase relationships combine in the image plane of an eyepiece to form a visible image.

The light which is diffracted by structural detail or discontinuities in the specimen is a quarter-wavelength out of phase with light which has not been diffracted. The direct light, with considerably greater intensity than the diffracted light, moves as a concentrated cone towards the ring of the diffraction plate. The relatively weak diffracted light, retarded in phase by a quarter-wavelength, moves so as to be distributed over the whole aperture of the diffraction plate (Fig. 3.5.3).

To compensate for the weaker diffracted light, a ring-shaped disc on the phase plate is

3.5.2
Stained specimens alter light amplitude.

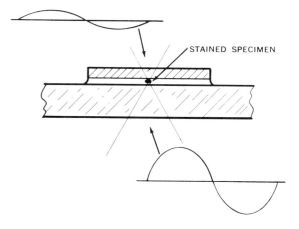

STAINED SPECIMEN

84

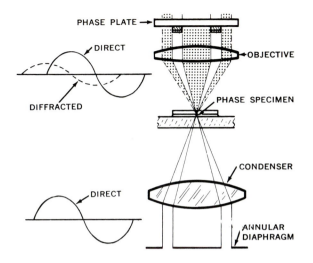

3.5.3
Unstained specimens diffract light so that it becomes a quarter-wavelength out of phase.

used to reduce the intensity of the direct light and equalise the brightness to achieve the desired contrast. In addition to this absorption, the diffraction plate uses a phase-retarding material so that the direct light is retarded by a quarter-wavelength, and this brings it into phase with the diffracted light. The resulting image of the specimen appears bright against a darker background (Fig. 3.5.4).

When the phase-retarding material is placed on all areas of the diffraction plate, other than the ring-shaped disc, a dark contrast image is produced because direct and diffracted light is brought to the image plane of the eyepiece a half-wavelength out of phase. A subtractive superposition of the light results in direct and diffracted waves cancelling each other to form an image darker than its surround (Fig. 3.5.4).

The standard working distance for intermediate phase objectives is 3 mm in air and, with a long-working-distance condenser it is 15 mm in air. One millimetre of air is equivalent to 1·33 mm of water or 1·52 mm of crown glass, and these ratios of refractive indices can be used to obtain the equivalent working distance when the specimen includes more than one medium (Fig. 3.5.5).

The use of plano–plano specimen preparations is of importance for phase microscopy and, when using intermediate or long working distances, the specimen should have optical-quality parallel walls with flat surfaces.

3.5.4
Bright-contrast phase image (*left*) and dark-contrast phase image (*right*).

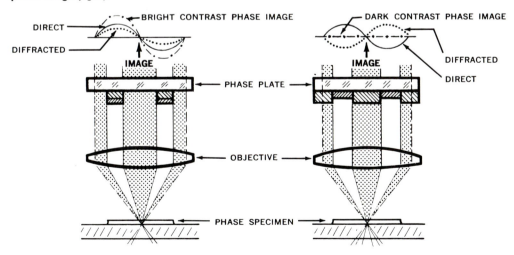

85

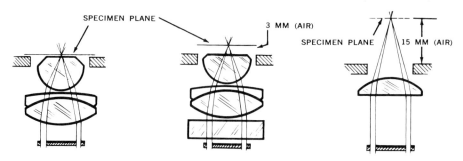

Standard Working Distance
Objectives: 10X, 20X, 45X, 100X

Intermediate Working Distance
Objectives: 10X, 20X, 45X, 100X

Long Working Distance
Objectives: 10X, 20X

3.5.5
Plano–plano specimen mounts are essential for phase microscopy.

Interference techniques

For basic materials research on crystals, metals, paints and plastics, and in industry for inspection, there is a need for the non-destructive measurement of surface finish characteristics such as length, slope and depth of surface features.

The surface finish interferometer, as fitted to the Vickers M41 Photoplan microscope allows visual estimation of surface deformation down to $\frac{1}{20}$ of the wavelength of mercury green light of 546 nm wavelength, representing about 27 nm steps. A higher accuracy is possible by the evaluation of fringe photographs to $\frac{1}{200}$ wavelength or 2·7 nm steps (Figs 3.7.1).

The microscopically enlarged surface of the specimen is crossed by a series of equidistant interference fringes, the spacing of which corresponds to a height change of exactly a half wavelength of mercury light through a 546 nm interference filter (Fig. 3.5.6).

The 40× objective is based on a modified

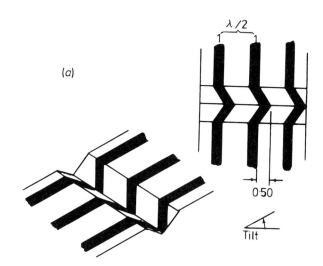

(a)

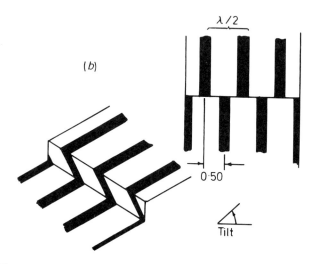

(b)

3.5.6
(a) Schematic appearance of fringes crossing a groove. The fringe deviation is 0·50 × fringe width; therefore, groove depth = 0·50 × λ/2 = λ/4. At wavelength 546 nm, groove depth = 136 nm. (b) Schematic appearance of fringe deviation at a surface step. Step depth = 0·50 × λ/2 = λ/4. This is 136 nm at 5·46 nm wavelength.

Mirau interferometer system. The reference surface is placed centrally on the objective front lens. The action of beam splitting is performed by a partially reflecting surface on a coverplate beneath the objective front lens. It will be seen that an illuminating ray first strikes the semi-reflecting surface where a proportion is reflected upwards illuminating the reference surface. After a further reflection from the semi-reflecting surface, the ray passes to the eyepiece. The other fraction of the light passes through the semi-reflecting surface to the specimen and then back up to the eyepiece. Destructive interference occurs when the two beams are 180° out of phase (Fig. 3.5.7).

The 10× objectives, based on the Michelson principle, contain a prism beam splitter beneath the objective front lens, together with a suitable reference surface. Interference occurs in exactly the same manner as with the 40× objective. Each type of objective is available, fitted either with a low-reflectivity reference surface for use with materials of low reflectance such as glass, dielectrics, etc, or with a high-reflectivity reference surface for use with materials of high reflectance such as polished metals and mirror coatings (Fig. 3.5.8).

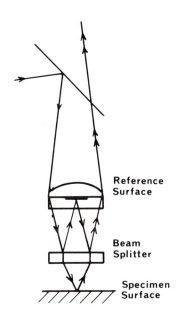

3.5.7
40× objective with modified Mirau interferometer.

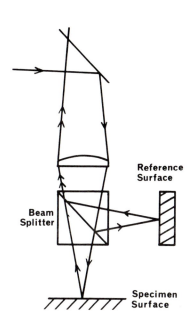

3.5.8
10× objective with Michelson-type interferometer.

3.6 Fluorescence microscopes

The luminescence phenomenon occurs in some substances which, if irradiated with ultraviolet, violet, blue or green light, emit visible radiation at a wavelength which is longer than that of the exciting light. If, after the irradiating light is switched off, the secondary emission persists for some time at an appreciable intensity, then the substance is said to be phosphorescent. If the excited radiation does not persist after the irradiating light is extinguished, then the material is fluorescent.

Some unstained specimens have the property of emitting fluorescent light when they are excited by short wavelengths, and this is known as primary fluorescence or autofluorescence. Generally, specimens need to be

87

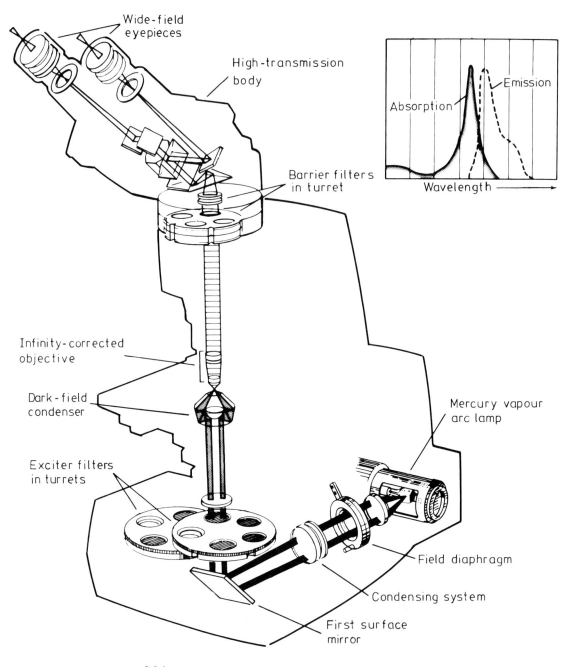

Wide-field eyepieces

High-transmission body

Barrier filters in turret

Absorption

Emission

Wavelength

Infinity-corrected objective

Dark-field condenser

Mercury vapour arc lamp

Exciter filters in turrets

Field diaphragm

Condensing system

First surface mirror

3.6.1
Schematic diagram of American Optical V20 'Fluorestar' microscope.

88

stained with fluorochrome dyes, and then tissues that are different will show up in contrasting colours, distinguished by fluorescence. The light emission is then known as secondary fluorescence.

Blue-light fluorescence is performed in the short-wave region of the visible spectrum around 400 nm wavelength by passing mercury light through a dark-blue filter. In order to observe the very weak fluorescent light, the blue exciting light which has passed through the specimen must be completely blocked out by a yellow-orange barrier filter (Fig. 5.5.1).

This technique permits observation of green, yellow and red fluorescent colours. Because the exciting light is in the shorter wavelength, a powerful tungsten–halogen light source may be used. A disadvantage of blue-light fluorescence is that blue-green primary fluorescence cannot be seen, as it will be absorbed by the yellow-orange barrier filter (Fig. 3.6.1).

Photomicrography of fluorescent materials is not easy because of the low emission of the light, so exposure times are long and difficult to determine. Owing to the comparatively low brightness of fluorescent images, photographs are usually recorded on high-speed 35 mm colour or black-and-white panchromatic film, and not on larger sizes of film or plates.

3.7 Polarising microscopes

Polarising filters are made of synthetic foil into which very small rod-shaped elements, known as micelles, are incorporated. When the foil is stretched the micelles are linearly oriented and act as a filter which transmits only one component of lightwaves and absorbs the other component. The emergent light is plane-polarised.

A polarising filter is fitted under the condenser of the polarising microscope and this is known as the polariser. Another filter, known as the analyser, is placed above the microscope stage and, if turned at right-angles to the polariser, it will not transmit any light. A polarising microscope is essentially a compound microscope to which polarising elements and a rotating stage have been added (Fig. 3.7.1).

The Vickers M41 polarising microscope is equipped for the systematic study of polarising effects which result when optically anisotropic

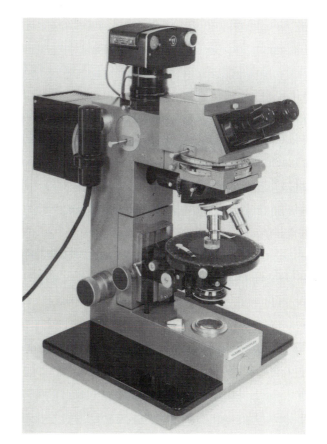

3.7.1
Vickers MH1 polarising microscope with incident illumination for examining anisotropic specimens in polarised light.

specimens are viewed in polarised light. To this end, it is provided with a precisely oriented polariser for illuminating the specimen with polarised light, with a graduated rotatable stage for placing the specimen in any required azimuth, and with an analyser for bringing the light components generated by the specimen birefringence into parallel polarised relationship. In addition, eyepiece cross lines are included to define the datum polarising directions of the polariser and analyser. Slots may be provided for the insertion of such birefringent compensators as quartz wedges which are used for the measurement of birefringence. More advanced instruments also provide a Bertrand lens for examining the interference figures formed at the back aperture of the objective in convergent light.

In the past, the polarising microscope has usually been associated with the studies of the mineralogist and petrologist, and for a long time the instrument was referred to as a petrological microscope. In recent years, however, it has been increasingly used in other fields for the analysis and identification of materials which are not easily identifiable by the usual chemical and physical methods, particularly when the samples are very small or their destruction not permissible. Today, in addition to the traditional fields of mineralogy and petrology, the polarising microscope is used in the fields of biology, chemistry, ceramics, brickmaking, pharmacy, metallurgy, textiles, civil engineering and soil mechanics.

4 Cameras for Industry and Commerce

4.1 Introduction

The *camera obscura* was the prototype of the modern photographic camera, and the earliest form consisted of a small hole in one wall of a dark room so that a sunlit scene outside cast an inverted image on an inside screen.

Over 900 years ago, Alhazen (962–1038) utilised the *camera obscura* to examine solar eclipses. His treatise on optics was translated into Latin by Witelo in 1270, then published by F. Risner in 1572, as the *Opticae Thesaurus Alhazeni Arabis*.

Giovanni Battista della Porta (1535–1615) gave details of the *camera obscura* in his book *Magia naturalis* and said that a lens should be used to form the image. The single meniscus photographic lens was originally designed in 1812 by William Hyde Wollaston (1766–1822) for use in a *camera obscura*. A *camera obscura* with mirror, lens and concave table viewing screen, for correcting field curvature, which was constructed in 1829 on Clifton Down, Bristol, overlooking the Clifton Suspension Bridge, is still in good working order and open to the public.

If the viewing screen is replaced with a photosensitive surface, then the *obscura* becomes a camera in the modern sense of the word. Photographic lenses are required to project an image on a flat plate or film mounted perpendicular to the optical axis. A good definition is needed for image points over the whole of the sensitive surface which may be 45° off-axis in some areas.

Joseph Nicephore Niepce (1765–1833) produced the first photograph in 1826 using a box camera, with a small convex lens, and paper sensitised with silver chloride. In 1835, William Henry Fox Talbot (1800–77) discovered that silver iodide on a paper negative would reduce the exposure time to half a minute from about one hour, and he invented the negative–positive process which has developed into photography as we know it today.

When cameras were first manufactured, the photographic lenses were single achromats based on John Dollond's invention in 1759 of the achromatic telescope objective. This type of lens was subject to aberrations and a secondary spectrum, which could only be corrected by using a small stop, so that the image was formed by rays passing mainly through the centre of the lens. One of the earliest camera portrait lenses was designed in 1840 by Joseph M. Petzval (1807–91) for Voigtlander and, in its original form, it had a speed of about $f/3.6$ (Fig. 4.1.1).

In 1891, the first telephoto lens was designed by Thomas Rudolph Dallmeyer (1859–1906) and this was followed in 1892 by

91

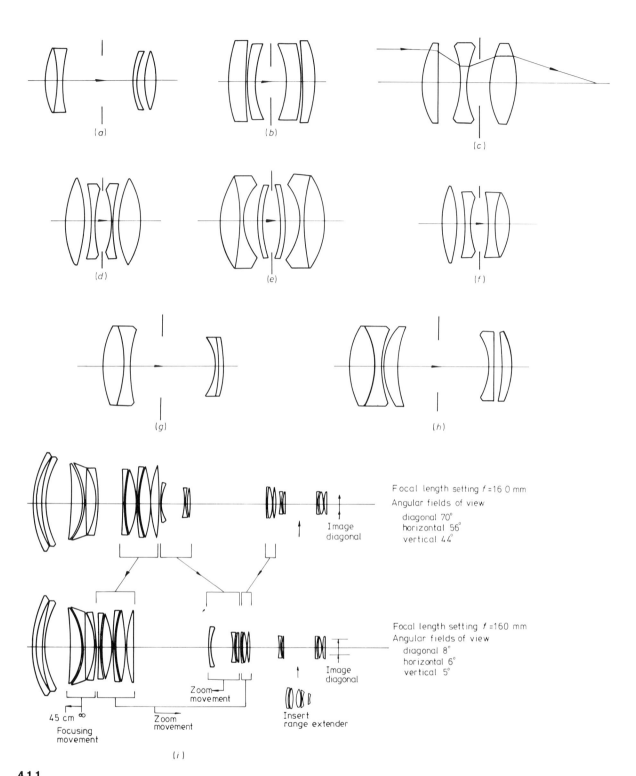

4.1.1

Typical photographic lenses. (*a*) **Petzval Portrait (1840).** (*b*) **The Gauss type of double lens, *f*/6·3 to *f*/4·5.** (*c*) **Cooke Triplet, *f*/3·5 (1893).** (*d*) **Cooke Aviar.** (*e*) **Ross wide-angle Xpres, *f*/4·0.** (*f*) **Zeiss Tessar.** (*g*) **Typical telephoto lens.** (*h*) **Distortionless telephoto lens.** (*i*) **Taylor–Hobson Varotal 30 TV zoom lens: camera format, 1¼ in plumbicon; range of focal lengths, 16–160 mm (0·63–6·3 in); maximum relative aperture, *f*/2·2; maximum transmission stop, *T*/2·5.**

92

the Adon telephoto lens. The anastigmat Cooke Triplet was patented in 1893 by H. Dennis Taylor (1862–1943) when he was employed by Cooke and Sons of York, and this design was manufactured under licence by Taylor, Taylor and Hobson Ltd. A very large number of high-quality lenses are now available for all types of cameras. The miniature camera, in many forms, dominates amateur photography and has been described in *Lens mechanism technology*.

The 35 mm miniature camera has several important advantages. Lenses of very wide aperture can be used which are not economic for larger types of camera. The depth of field is a function of lens diameter so, for a given aperture, the short focus lens has an advantage after making allowances for different sizes of image.

Disadvantages of miniature cameras include the need for extreme care in processing the film as blemishes and faults are relatively more important. Negative retouching is practically impossible and great care is necessary to avoid camera shake, so short exposures, on high-speed film, are essential using objective lenses of large relative aperture.

Very fine grain 35 mm film is now available but, because enlargement of the small images by projection is essential, the grain size will always require careful control. The resolving power of the camera depends on the silver halide photographic emulsion as well as on the objective lens. Overexposure and over-development increase graininess and grain size also increases with the sensitivity of the film. Because grain size and sensitivity are related, the response of an emulsion to a range of light intensities depends to some extent on the size distribution of the silver grains. The silver halides themselves are only sensitive to blue radiation but, by the addition of dyes, the sensitivity can be extended throughout the visible spectrum into the near-infrared.

Richard Leach Maddox (1816–1902) in 1871 first successfully used gelatine as the basis for a silver bromide emulsion. Gelatine acts as a halogen acceptor and, when light acts on the silver bromide, the liberated bromine is taken up by the gelatine and reversal after exposure is prevented.

Reflex cameras are widely used by press and professional photographers who wish to use a large negative and so obtain the highest possible performance. The $2\frac{1}{4}$ in $\times$ $2\frac{1}{4}$ in (57 mm) square format gives a better quality than the 35 mm size, but is small enough to be reasonably light and portable. The single-lens reflex camera ensures a complete absence of parallax, and interchangeable backs permit a rapid change of film from black-and-white to colour. Some reflex cameras use a focal-plane shutter, and others employ a between-the-lens shutter. There are formidable design problems to overcome so that the film is protected from light whilst the image is being focused through the lens onto a ground-glass screen.

The advantages of this arrangement are that the picture is seen at full brilliance over the whole film area and, by adjusting the rise and fall or sideways movement of the camera front, the subject can be arranged to give best results. The focusing is easy and exact up to the instant of exposure irrespective of the focal length or aperture setting to be used.

When using a rising-front camera it is essential to use a wide-angle lens as the displacement of the lens, relative to the centre of the negative, requires that is must cover a larger than normal field.

In photography, the relation between the subject, the lens and the image is given by

$$\frac{1}{f} = \frac{1}{v} + \frac{1}{u}$$

where f is the focal length of the lens, and v and u are the distances between image and lens and subject and lens, respectively.

In aerial photography, u may be considered to be infinite compared with f and v, in which case the formula becomes

$$v = f.$$

It can therefore be assumed that the image is always formed in the focal plane of the lens, and for this reason all cameras used for aerial photography are arranged so that the sensitised surface of the plate or film is exposed in

this plane; in other words, aerial cameras are of fixed focus.

Colour process lenses must be very highly corrected for lateral chromatic aberrations and for red, green and blue longitudinal chromatic aberrations. Apochromatic lenses usually have a long focal length which helps in the achievement of good definition over large plate sizes for half-tone, line and colour-separation negatives.

Copying lenses are special types of enlarging lens which have been designed for a high performance over restricted conditions, such as use at only one or a small range of magnification ratios (Fig. 4.1.2).

4.1.2
Copying lenses are designed for use with object points which lie relatively close to the lens. The aberrations are balanced for this condition and may not be balanced for distant object points.

4.2 Motion-picture cameras

35 mm motion-picture cameras

Motion pictures consist of a large number of 'still' pictures which are photographed in rapid succession. A rate of 16 picture frames per second is adequate to give an impression of continuous motion, due to persistence of vision, but a rate of 24 frames per second is necessary for sound-on-film in order to provide good audio reproduction. At 24 frames per second, the 35 mm wide film will pass through a camera at a rate of 90 ft min^{-1} (27·432 m min^{-1}) or 18 in s^{-1} (457·2 mm s^{-1}) per second. In order to synchronise with the 50 Hz electricity supply, all motion pictures used for television transmission are photographed at 25 frames per second. For slow-motion effects, the cameras are designed to operate at speeds up to 100 frames per second.

Arnold and Richter KG, München, is one of the leading manufacturers of professional motion-picture cameras which are used for photography in real locations and sporting events. The Arriflex 35BL has a film-transport system, utilising two dual-transport claws and two registration pins, with counterbalanced low-mass components to ensure vibrationless operation. The two registration pins hold the film stationary during the exposure cycle and a fixed gap channel, with a small back-pressure pad, prevents emulsion build-up in the gate (Fig. 4.2.1).

Thickness of photographic film stock varies, but it is generally accepted as being 0·006 in (0·15 mm) with the film base as 0·004 in (0·10 mm) and the emulsion 0·002 in (0·05 mm). Aperture plates can be changed to suit the aperture sizes required for various projection formats (Fig. 4.2.2).

The viewfinder eyepiece, containing ground glass marked with the desired format, can be removed without influencing the adjustment between the film image and the viewfinder image. During film transport, the shutter wing, entering the light beam from the taking lens, reflects the image produced by the lens into the viewfinder without any loss of light. A magnified image visible through the

4.2.1
Arriflex 35BL mirror-reflex motion-picture camera for location synchronous-sound filming.

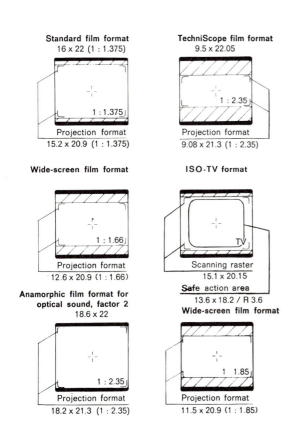

Standard film format
16 x 22 (1 : 1.375)

1 : 1.375

Projection format
15.2 x 20.9 (1 : 1.375)

Wide-screen film format

1 : 1.66

Projection format
12.6 x 20.9 (1 : 1.66)

Anamorphic film format for optical sound, factor 2
18.6 x 22

1 : 2.35

Projection format
18.2 x 21.3 (1 : 2.35)

TechniScope film format
9.5 x 22.05

1 : 2.35

Projection format
9.08 x 21.3 (1 : 2.35)

ISO-TV format

TV

Scanning raster
15.1 x 20.15

Safe action area
13.6 x 18.2 / R 3.6

Wide-screen film format

1 : 1.85

Projection format
11.5 x 20.9 (1 : 1.85)

4.2.2
Film formats.

eye-piece, whether the camera is running or at a standstill, is free of parallax and identical in focus with the film image (Fig. 4.2.3).

The camera can be used with anamorphic lenses for Cinemascope films. Alternatively, hand-operated zoom lenses can be fitted or a servo drive used for fast or slow uniform adjustments to the lens focal length. (For further details of the Angénieux zoom lens see Chapter 2.3 of *Lens mechanism technology*.) Interchangeable magazines can contain 400 ft (120 m) or 1000 ft (300 m) of film (Fig. 4.2.4).

16 mm motion-picture cameras

At a camera speed of 24 frames per second, the 16 mm film travels at 36 ft min^{-1} (10·97 m min^{-1}) and, because this is only 40% of the 35 mm film speed, the magnetic sound frequency recorded will be only about 5 to 8 Hz compared with 10 to 15 Hz for 35 mm film.

95

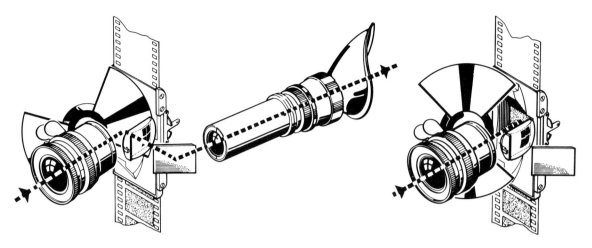

4.2.3
**The mirror-reflex viewing system. A double-blade shutter
rotates at a 45° angle between the taking lens and the
optical axis. This rotary shutter—the heart of the
mirror-reflex system—is made of special glass with a
mirrored surface, most accurately balanced, and with
maximum optical precision.**

4.2.4
**Arriflex 35BL motion-picture camera with Angénieux
25–250 mm zoom lens on support brackets and magazine
for 400 ft (120 m) of film.**

96

The Arriflex 16ST camera has a 180° rotary mirror-reflex shutter, which makes one complete turn for every two frames, and so all the light entering the lens is available alternatively for viewing and filming. The shutter opening results in the following exposure times.

Camera speed (FPS)	Exposure (s)
8	1/16
24	1/48
32	1/64
48	1/96

The exposure can be calculated by doubling the speed figure and reading the result as a fraction of a second (Fig. 4.2.5).

The combination of transport claws and pilot pins in the film-drive mechanism provides a steady image. The transport claw enters the film perforations from the lens side of the film and the pilot pin from the other side. This ensures high-quality duplicate prints whether 16 mm to 16 mm, 16 mm to 35 mm, or 16 mm to 8 mm are required (Fig. 4.2.6).

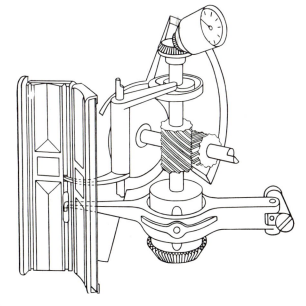

The ground-glass image in the viewfinder is viewed through a 10× magnifying eyepiece which features an automatic closure mechanism. This opens automatically when pressure is applied to the rubber eyecup and closes when pressure is removed, and so accidental film fogging through the viewfinder is prevented (Fig. 4.2.7).

The cadmium selenide exposure-control system measures light behind the lens in a central circular area equal to 50% of the frame aperture. All factors that influence exposure, such as light transmission of the lens, f stop, lens extension, lens field angle and filters, are taken into account. The light-sensing elements are in the viewfinder system, and not in the filming system, so there is nothing between the lens and film to scatter light or degrade picture quality (Fig. 4.2.8).

The Arriflex 16BL is equipped with synchronous sound recording for a location double system and a synchronising signal from the camera is carried to a tape recorder by connecting cables and recorded on the tape-head. Marks on the film and sound-track are

recorded each time the camera is started, and so location sound filming is possible in perfect synchronisation. The 12 V D.C. universal motor is controlled by a crystal oscillator, which provides a reference frequency, and a constant speed is achieved. Alternatively, if

4.2.7
Mirror-reflex viewing system. Light beam during the film transport phase.

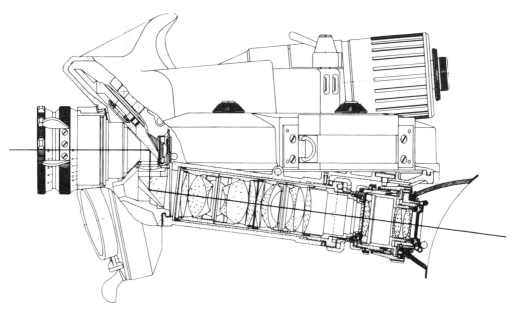

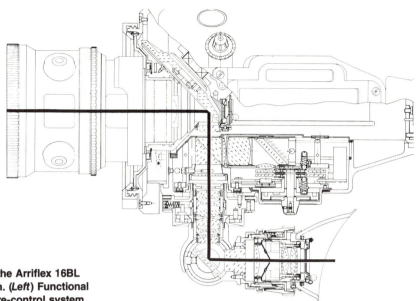

4.2.8
(Above) The mirror-reflex system of the Arriflex 16BL with built-in exposure-control system. *(Left)* Functional schematic of the 16BL CdSe exposure-control system.

electric power is available, the motor speed can be regulated with reference to line frequency.

Many films can be made to advantage by a single-system sound if there are long uninterrupted takes and simple production techniques such as for television, public relations, and training. A single-system sound module includes the magnetic record and separate playback heads, a massive flywheel to eliminate wow and flutter, and essential film guides. The magnetic sound record is displaced 28 frames ahead of the picture frames (Fig. 4.2.9).

4.2.9
Arriflex 16BL light-weight reflex sound camera with fast threading of the film over magnetic soundheads.

4.3 Sound-recording cameras

Sound recording on magnetic tape has been used for original recordings since the early 1950s but, because of the need for a permanent record, photographic sound-on-film (which was first produced under commercial conditions in 1927) is still used for release prints to be used for reproduction in cinemas or by television. The original sound-track, and dubbing or laboratory mixing, is recorded on magnetic tape but the final release prints are recorded on fine-grain emulsion photographic film.

The sound-track on a 35 mm film is printed at a distance of 19 frames, or approximately 355 mm, ahead of the corresponding picture. Separate cellulose acetate based films are used for picture and sound negatives, whilst the 19 frame displacement is set when making the

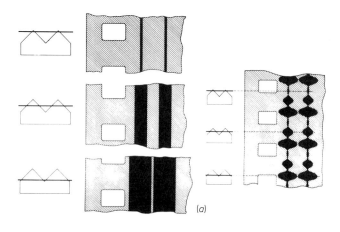

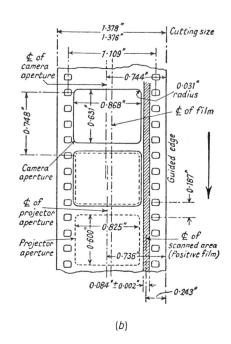

(b)

4.3.1

(a) Variable-area sound-track. (b) Layout of 35 mm sound film and apertures. (c) Dimensions for 35 mm negative and positive sound-track. (d) Dimensions for 16 mm negative and positive sound-track.

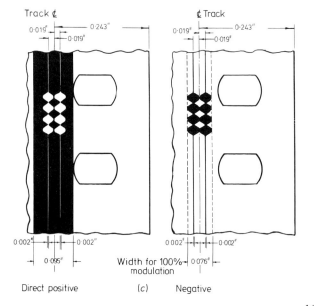

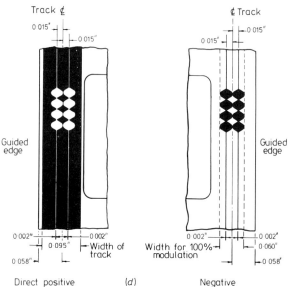

combined release print.

There are two methods of producing a photographic record of sound-on-film, by tracks of variable area or variable density. The speed of the moving film at the point of recording is exactly 90 ft min^{-1} for 35 mm film and 36 ft min^{-1} for 16 mm film (Fig. 4.3.1).

The optical section of the variable-area recorder consists of a mirror galvanometer, including a series of lenses and masks, the lamp and the film. A very small mirror is mounted on tension strips across the poles of a high-flux permanent magnet. Between the pole pieces, a thin armature vibrates in accordance with the modulation of the magnetic field caused by the audiofrequency currents in the armature coil. One end of the armature engages in a groove at the back of the mirror holder (Fig. 4.3.2).

The triangular spot of light reflected from the galvanometer mirror moves up and down across the second mask and so illuminates more or less of the slit in the mask. This variable length on the slit controls the width of

4.3.2
Optical sound-recording camera. The film is exposed in the central dark compartment and the mirror galvanometer is contained in the right-hand side of the recorder. Unexposed film in the left-hand magazine passes through the camera mechanism to the right-hand magazine.

the sound-track and the time intervals between successive peaks indicates the sound frequency (Fig. 4.3.3).

The variable-area sound-track produced is of double-bilateral form resulting from the use of a twin-triangular mask in the image-forming optics. This type of track has the advantage that at low modulation levels the signal waveform is near the centre of the track where, in reproduction, there is least possibility of distortion from non-linearly.

In the variable-density system, the recorder consists of a light valve with an opening defined by metal ribbons. Amplitude variations result in the closing and opening of these ribbons, and the maximum opening is twice the width of the unmodulated opening. Correct sound reproduction depends on the image densities, and it is necessary to develop the negative so that the product of the negative-film γ and the positive-film γ will be unity when the various correction factors have been applied.

The unexposed film is stored in the left-hand section of the detachable magazine above the mechanism. The film passes round a drum in the camera, where the recording light beam from the slit illuminates and exposes the film, to the right-hand section of the magazine. In order to record high-quality sound, the film speed must be exactly 90 ft min^{-1} and this is achieved with a specially designed electromagnetic flywheel.

4.3.3
Path of light through the optical parts of the recording machine.

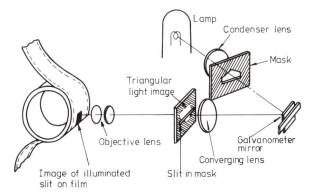

101

It is customary to record only the important sounds whilst photographing the action and to add noise effects or background music in a subsequent re-recording process. Another method is to project the film in a small 'dubbing' theatre and produce additional sounds from microphones.

When recording on negative film is completed, the picture and sound images must be transferred to positive film by a printing process. Usually, the positive and negative films are made to come into contact on a large-diameter sprocket wheel or curved gate and in the middle is an exposing light source. The diameter of the sprocket or curvature of the gate is designed so that the difference between the radii of the negative and positive arcs will allow for average shrinking of the negative.

Because of variations in average negative density from scene to scene, and to ensure a balanced print, the original negatives are graded and the printer must incorporate a system for variations in the light for exposure of the positive print. Usually, the light changes are effected by means of a shutter, which moves across the printer gate, controlled by notches which have been previously cut in the edge of the negative.

4.4 High-speed cameras

Some 35 mm motion-picture cameras can operate at speeds up to 100 pictures per second, and specially modified 16 mm intermittent cameras have been designed for speeds up to 500 pictures per second.

Applications for high-speed cinematography, in scientific and technical investigations, have increased from 50 000 pictures per second in the 1950s to 8 000 000 per second in the 1970s, and this has been made possible by the development of a rotating-mirror camera. Studies concerning the explosion of a mining detonator, the containment of plasma during fusion, the emission of a laser, the development of a crack in glass or brittle metal, pre-ignition in an internal combustion engine or the collision of a raindrop on the wing of a supersonic aircraft all require high-speed cameras.

Film moving through a camera at about 100 m s^{-1} will provide 44 000 pictures per second, from a rotating glass block, but the film is not strong enough to go any faster. Ultra-high speed demands a stationary film and the image of an event reflected from the surface of a small mirror rotating at very high speed (Fig. 4.4.1).

The Barr and Stroud framing camera CP5 records an event as a sequence of frames or pictures, but it can be converted to provide a continuous streak image. In the framing mode, the light from an event passes through an objective lens, and shutter assembly, to be focused on a stainless-steel rotating mirror. The light is reflected from the mirror and passes through a series of lenses which relay the image on 35 mm high-speed emulsion film to produce a sequence of circular pictures each 8·2 mm diameter. At the maximum operating

4.4.1
Barr and Stroud ultra-high-speed camera for photographing transient phenomena by utilising a rotating mirror.

102

speed of 8.0×10^6 pictures per second, the mirror is rotated by an air-driven turbine at a speed of 333 000 revolutions per minute (rpm) and 117 frames are exposed in a time of 0·12 μs per frame. The stainless-steel mirror is 27 mm in diameter, polished on both sides, and coated with a thin film of aluminium protected with silicon dioxide. It rotates in a vacuum housing at a pressure of 5 mmHg (Fig. 4.4.2).

A camera of this design has an intrinsically low optical aperture from $f/13$ to $f/16$ and can only take pictures when the mirror is reflecting images onto the arc lenses. The camera

is therefore 'blind' for a large part of the time, and so the event has to be timed to coincide with the moment when the camera can record the images. Each secondary lens will produce an image on the film whilst the light reflected from the mirror passes through it.

The images recorded on the film are separated in time by an amount directly proportional to the rotation speed of the mirror. The timing sequence is automatic and, when the mirror reaches its preset speed, light from a synchronising lamp is reflected by the rotating mirror onto a photodetector and the amplified output triggers a delay unit to start the event.

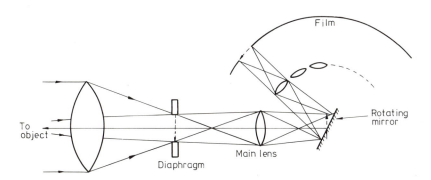

4.4.2
The objective lens forms an image of the object and this image is then focused by the main lens on the rotating mirror. The diaphragm is placed in the plane of the image formed by the objective lens and serves to limit the angle of the marginal rays of light in the system thereafter, so as to prevent more than two of the secondary lenses from being illuminated at any instant.

4.5 Cameras for professional and aerial photography

Action photography, such as press and sports photography, calls for a lens aperture as wide as possible, whereas architecture and advertising require the use of camera movements, and a large covering power by the lens is of primary importance. Tripod time exposures of stationary subjects will achieve the best results if a large camera format and adequate lens coverage permits scope for lens displacement.

The Linhof Master Technika is a professional camera in 4 in × 5 in (9 cm × 12 cm) format which requires very wide-angle lenses in order to make use of its extreme adjustment facilities (Fig. 4.5.1).

The depth-of-field range can be expanded so that, even with subjects located obliquely at different distances from the camera, the advantage of short exposure times on colour photographs can be exploited without stopping down (Fig. 4.5.2).

4.5.1
Linhof Master Technika in 4 in × 5 in format.

103

4.5.2

(a)–(b) Examples of use of the Linhof Master Technika.

Basic position

(a)

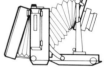
Lens standard in elevated position. Eliminates converging lines in architectural and industrial photography when shooting from a low camera position.

Basic position

(b)

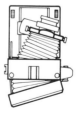
Camera adjustments according to the Scheimpflug rule. Increased depth of field, in line with the subject's requirements, in the horizontal plane. Ideal for close-ups.

Basic
position

Lowering the op-
tical axis by drop-
ping the lens stan-
dard. This technique
renders vertical
lines parallel when
shooting from an
elevated position.

(c)

Basic
position

Camera adjust-
ments in vertical
position according
to the Scheimpflug
rule. To extend
depth of field from
the foreground to
infinity.

(d)

105

4.5.3
(a) The Linhof Electric 70 is a development of the Aero Press camera for aerial photography, recording and series documentation. (b) The Linhof Aero Electric has the film transport and shutter operated by an electric motor with power supply from batteries. Motion sequences are recorded from a solenoid remote release.

Tapered bellows permit large adjustments of the lens and film holder. As it is customary to compose the picture on a ground-glass screen, fitted with a Fresnel lens to provide uniform brightness, before the film is placed in position, the multi-focus viewfinder is only needed when the camera is used for action photography.

The Linhof Aero Electric camera was developed from the Aero Press camera and is designed for aerial photography, but it is also suitable for documentation photographs and news reporting. Time-lapse photography is also possible by remote control through cables or by radio (Fig. 4.5.3).

The camera is electrically operated with a 24 V power supply to the film-transport and shutter mechanisms (see section 4.4 of *Lens mechanism technology*). All lenses are mounted in Synchro-Compur shutters and are suitable for the negative size of $2\frac{1}{4}$ in $\times$ $2\frac{3}{4}$ in (56 mm $\times$ 72 mm) (see section 7.1).

The Linhof Aero Technika is designed specifically for oblique aerial photography in the 4 in $\times$ 5 in format. The roll-film vacuum holder, with motor-driven transport, allows rapid sequences of up to 150 exposures per magazine loading in 5 min as the sequence time is no more that 2·5 s (Fig. 4.5.4) (see section 7.1).

4.5.4
The Linhof Aero Technika with 4 in $\times$ 5 in format has an electromagnetic shutter and motor-driven film transport.

4.6 Document copying

The first known copying machine was invented in 1780 by James Watt, who tried to make an offset printing machine for copying correspondence. In 1842, the blueprint process (white lines on blue background) was developed for architectural and engineering drawing reproduction. A direct photographic reproduction process, known as the 'Photostat', was introduced around 1900, but this was expensive both in equipment and sensitised materials and so had only a limited use but was available until about 1940.

In the 1930s, the diazo copying process was developed, for use in drawing offices, in which a translucent original has to be placed face upwards over coated diazo paper. The image was transferred to the diazo paper by ultraviolet light and developed with ammonia. The diazonium salts are much cheaper than the silver salts required by the photographic process, but the process was not suitable for the office copying-machine market.

Although scientists had observed electrostatic phenomena centuries before dynamic electricity was understood, there was no practical application of electrostatic recording before 1900 and no commercial equipment available before the 1930s.

In 1934, Chester F. Carlson, a physicist who had become a patent attorney, set out to find or invent a new process which could form the basis for an office copying machine. Carlson, using the principles of static electricity, had the idea that if a photoconductive material could be formed into a thin layer it would serve as a photographic plate in which electric currents could be controlled by a light pattern or image, and in 1937 he filed a preliminary patent application.

After further experiments, Carlson, in 1940, approached some 20 companies in America to try to obtain commercial support for the invention, but without success. However, the Battelle Memorial Institute of Columbus, Ohio, agreed to undertake further development of 'electrophotography' and larger-scale research was started.

In a patent issued in 1944, Chester F. Carlson proposed the use of an electrostatic printing plate consisting of insulating characters bonded to a conductive metal sheet. By frictional charging, the characters could be made to pick up a pigmented resin powder and the powder image was then transferred to a moistened or heated sheet of paper which was then passed through a fuser to fix the image. This was the basic patent following which an automatic copying machine was built and tested with success (Fig. 4.6.1).

In 1947, development was accelerated by the Haloid Company of Rochester, New York (now Xerox Corporation), acquiring a licence and supporting research at Battelle. At that time, the process was named 'Xerography' from the Greek *xeros* meaning 'dry' and *graphos* meaning 'writing' to emphasise the totally dry nature of the process as opposed to wet photographic methods.

Nearly all phases of the process and equipment were refined and improved at Battelle so that the first commercial flat-plate Xerox Copier was on the market in 1950. Since 1950, xerography using selenium as the photoconductor, has been developed to a high state of perfection.

In May 1955, xerography was demonstrated in Europe and, in November 1956, Rank Xerox Ltd was formed to market xerographic equipment manufactured by Rank Precision Industries Ltd (Fig. 1.7.6).

The xerographic image can be either:

(*a*) A high-contrast pattern on an insulating film.

4.6.1
Combination of photography and printing.

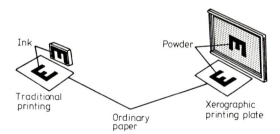

107

(b) A continuous-tone pattern on an insulating film.

(c) A pattern of insulating material on a conductive base.

The vitrous selenium xerographic photo-receptor has a sensitivity corresponding to an ASA exposure index of 3 and is sensitive at wavelengths from 280 nm, in the ultraviolet, to about 550 nm, but this range can be extended into the near-infrared by the addition of small amounts of tellurium. Selenium photoreceptors copy blue lines on white paper with difficulty, because the blue image discharges the electrical image of the predominantly blue-sensitive selenium almost to the same extent as does the white background on which it appears (Fig. 4.6.2).

Selenium plates and drums consist of chemically cleaned, uniformly oxidised aluminium onto which high-purity amorphous selenium is vacuum-evaporated to a thickness of 20 to 160 μm. When flat sheets are required, the necessary surface quality is obtained by special rolling techniques, whereas cylindrical drums are diamond-turned and the finish must be as near perfect as possible without any isolated defects. Heat treatment of the aluminium substrate forms an interfacial barrier of aluminium oxide between the base

electrode and the selenium storage layer. The photoconductive layer of selenium in the vitreous allotrope is formed by condensation of the vapour in a vacuum of less than 10^{-4} mmHg.

The electrical potential required across the photoreceptor layer to sensitise it is usually by the corona discharge at a fine wire (Fig. 4.6.3).

The exposure of the photoreceptor, to a subject in the dark, results in the formation of an electrostatic charge pattern, and the difference in potential between exposed and unexposed areas is called the electrostatic contrast, and this determines the development of the image.

Normally, electrostatic images are made visible by bringing close to them a pigmented, finely divided powder which is attracted to the electrostatic image. For better control of development the powder or 'toner' is mixed with a much coarser granular material known as the 'carrier' and the combination is cascaded across the exposed xerographic photoreceptor.

Two-component developers use the carrier to charge the toner by contact electrification; this is also called triboelectrification from the Greek *tribein* meaning 'to rub'. The image particles consist of extremely fine carbon-black powder dispersed in a thermoplastic

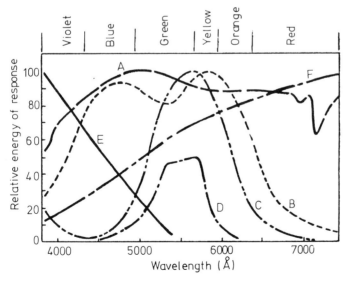

4.6.2
Relative spectral energy of typical emitters and response of typical xerographic photoreceptors. Adopting the nomenclature of conventional photography, the difference in potential between exposed and unexposed areas is called the electrostatic contrast. The curves represent: A, daylight; B, daylight fluorescent; C, eye sensitivity; D, Rose Bengal dyed zinc oxide; E, vitreous selenium; F, photoflood (3360 K).

108

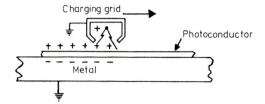

4.6.4
Charging the selenium-coated drum.

polymer which must be easily heat-fused. The 'carrier' is a special sand, or may consist of glass beads, and must be of a carefully controlled size to give high resolution without blocking or adhering to the selenium plate because of strong local fields.

The toner carbon-black particles average 6 µm diameter, whereas the carrier sand particles or glass beads are about 700–750 µm diameter. The toner is negatively charged on the carrier and, as the toner is stripped away during development, the net charge remaining must be removed by a grounded metal surface.

Amorphous selenium, which is a supercooled liquid, has a high resistivity which enables development to be completed before the electrostatic image is destroyed by electrical conduction through the plate. The process consists of:

(a) Electrostatic sensitising or charging.
(b) Image exposure.
(c) Electrostatic image storage.
(d) Image development.
(e) Image transfer to paper.
(f) Regeneration processes.

The electrostatic image is a positive charge pattern, on the photoreceptor surface, which corresponds to the illuminated pattern to be printed. Each toner particle must bear sufficient electric charge so that the electrostatic image fields will capture the particle from the

mass of the developer. If the charge on the toner is too great, the latent image charges will be neutralised before the development is dense enough for printing.

As the temperature of the toner is raised, for heat fixing on the paper, the particles wet each other and coalesce or flow together and then wet the paper surface and penetrate into the paper fibres to form a permanent bond. The first operation is to charge the surface of a revolving selenium-coated drum with static electricity (Fig. 4.6.4).

An image of the document to be copied is projected through a lens onto the charged drum. Because of a property of the selenium coating on the drum, the charge disappears from all areas where there is no writing on the original. Where there is writing, the drum has on its surface an exact copy of the original in electrostatic charge, which is of course invisible (Fig. 4.6.5).

4.6.5
Projecting the document to form a latent electrostatic image on the drum.

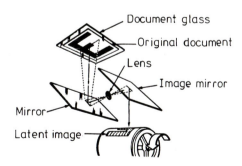

The developer containing toner is now poured over the drum. This powder also has an electrostatic charge, but opposite to that on the drum. The opposite static electrical charges on drum and toner are attracted, and therefore the drum now has an image of the original in toner powder (Fig. 4.6.6).

Paper is now placed under the drum after the toner has been poured on. Underneath the paper is placed a very large charge of static electricity of the same type as that on the drum. This attracts the toner off the drum and onto the paper, and leaves an exact copy of the original on the paper (Fig. 4.6.7). The paper is now passed under a source of heat which melts and fuses the toner to the paper to give a permanent copy of the original (Fig. 4.6.8). The drum now continues to a point where it is cleaned by neutralisation with an opposite charge, the odd pieces of toner which may be left being brushed off with a soft brush or web (Fig. 4.6.9).

Although there are many ways of designing the projection head for reproduction of micro-film or characters from a cathode ray tube, engineering drawings, or documents of various sizes, the basic camera and recording mechanisms are very similar on all models (Fig. 4.6.10).

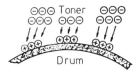

4.6.6
An image is formed in toner powder on the drum by electrostatic attraction.

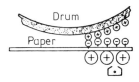

4.6.7
Transfer of image from drum to paper.

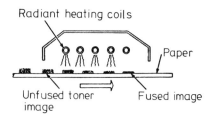

4.6.8
Fusing the toner image on the paper under a heat source.

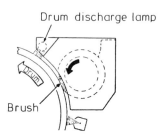

4.6.9
Cleaning the drum.

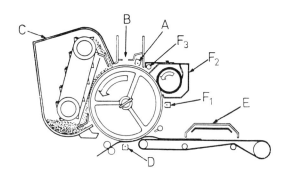

4.6.10
Typical xerographic mechanism: A, charge corotron consisting of 0·003 in platinum and rhodium wire at about 8000 V; B, image of document projected onto the drum; C, the developer box containing carrier and toner; D, transfer corotron to place a positive charge under the paper which is greater than the positive charge on the drum; E, at D the toner is transferred to the paper where it is fused and made permanent under radiant heat; F_1, the pre-cleaned corotron gives a negative charge which discharges the drum; F_2, the drum is rubbed with a fur brush to remove the latent image; F_3, a strip light removes all traces of the static charge before recharging again by the corotron.

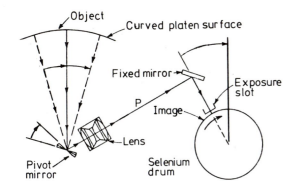

4.6.11
Schematic diagram of optical system in the Xerox Model 2400 copier, showing the principal optical ray (P), and the scan arcs of the principal optical ray, pivot mirror and selenium drum.

Xerox Models 2400 and 3600 with a maximum image size of 8½ in × 13 in (21·6 cm × 33 cm) had a common optical system and, with improved selenium sensitivity, the 3600 reproduced 60 prints a minute (Fig. 4.6.11).

With permission of Rank Xerox Ltd and the Society of Photo-Optical Instrumentation Engineers, who published 'M.T.F. in Lens design by Xerox' by George L. McCrobie in their proceedings, some details are given of the optical layout in Xerox Model 9200 which includes a zoom lens and photoreceptor belt instead of a drum. From a rear view of the lens, the motor-driven cam shaft can be seen. The cam followers vary the relative positions of the rear doublet lens and the rear outer component. A third cam moves the front outer component and the main ramp travel system of the entire lens (Fig. 4.6.12).

With further improvements in sensitivity of the photoreceptor, and the introduction of a belt instead of a selenium-coated aluminium drum, xenon flash illumination of the document has replaced the strip lighting of earlier

4.6.12
Zoom lens for Xerox Model 9200.

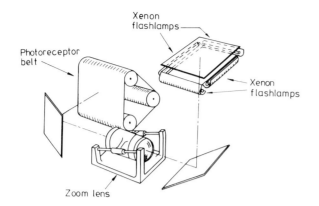

models and so, by eliminating the slit, the whole coverage of the lens can be used. The demand for varying sizes of copies has been achieved by the use of a very high-quality finite conjugate zoom lens (Fig. 4.6.13).

4.6.13
Xerox Model 9200 optical layout using flash exposure and a photoreceptor belt.

4.7 Letterpress and lithography

Printing is a very ancient art but, in Europe, it can be said to date from the invention of movable type, as previously the relief letters were carved in the surface of wood. The three methods of printing in general use are letterpress or relief printing, lithographic or planographic printing, and gravure or intaglio printing, in which the printing surface is recessed (Fig. 4.7.1).

4.7.1
The three methods of printing.

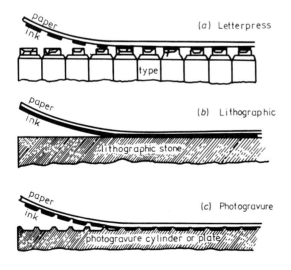

Letterpress

The invention of paper is attributed to Tsai Lun in China in about A.D. 105. The first reliable record of a paper-mill in Germany was at Nuremburg by Ulman Stromer in 1390. Printing with discrete type existed in China and Korea at about the middle of the eleventh century. This type was made from baked clay, but the printing of textiles, playing cards, and religious pictures with carved wood-cut blocks dates back to the sixth century A.D.

Johannes Gutenberg (1397–1468) of Mainz was the inventor of printing, about the year 1450, using renewable metal type. The first books to be printed with movable type, about 200 copies of the Gutenberg Bible, were completed in 1455. The letters and symbols were produced by pouring molten metal, consisting of an alloy of lead, tin, antimony and a small amount of bismuth, into a hand casting die. William Caxton printed the first book in England during 1476, and the first illustrated book *Myrrour of the Worlde* in 1480.

In 1884, Ottmar Mergenthaler (1854–99) completed the first machine for mechanical composition of letterpress type under the name of 'Linotype'. This was followed in 1887 by Tolbert E. Lanston (1844–1913) who automatically cast one individual letter at a time onto engraved brass matrices with a 'Monotype' machine. The letterpress method

112

may be duplicated easily by stereotyping using a flexible form which can be used for casting curved plates ready for use on rotary cylinder presses.

Photographs to be used for illustrations must be rephotographed through a half-tone screen with a cross-line ruling, depending on the quality of the paper, from 65 lines/in (25 lines/cm) for newspapers to 133 lines/in (52 lines/cm) for commercial work on smooth-coated art papers.

A glass half-tone screen is made by ruling lines through a resist on two optically flat pieces of glass. The lines are then etched into the glass and filled with black pigment. The two pieces of glass are cemented, so that the lines are at right-angles to each other, and the edges bound together (Fig. 4.7.2).

Half-tone screens which are used in process cameras must be large enough to take photographic films of the same size as the printed page images. The original photograph is placed on the copy board of a process camera fitted with a right-angled prism and plane mirror to give the lateral reversal which is essential for letterpress blocks (Fig. 4.7.3).

The short distance between a glass screen and film is critical in order to obtain a correct dot structure for tone gradation and contrast. Alternatively, contact screens can be used which are clamped by vacuum against the photographic film. Contact screens are made of cellulose acetate material, coloured either magenta or grey, and the emulsion side of the screen must be against the emulsion side of the film. The magenta contact screen is used with black-and-white negatives and for making half-tone positives from colour-separation negatives. Grey screens are used for direct colour-separation work. Kodalith Autoscreen orthochromatic film, which is sensitive to blue, green and yellow light, when exposed to a continuous-tone original produces a dot pattern automatically without the use of a screen. This film can be used in any camera fitted with a filmholder as no vacuum back is necessary. Process cameras are made in many sizes for sheet or roll film, some are of horizontal design and others vertical in order to save floor space (Figs 4.7.4 and 4.7.5).

The camera is fitted with tapered bellows so that the distance from the filmholder to the lens and from the lens to the copy board may be varied in order to produce an enlargement, same size, or reduction. When setting for same size, the distance between the copy and the film must be four focal lengths, with the nodal point of the lens two focal lengths from the copy. Half-tone photography, and the production of small closely spaced dots, when reproduced by printing gives the optical illusion of continuous tone at a distance from the eye.

The film, produced after photographing the illustration and text through a screen, is then contact-printed on a zinc or copper plate which has been treated with a light-sensitive colloidal substance such as bichromated

4.7.2 A glass screen is placed in front of the film inside the process camera. The screen (shown enlarged on the right) is formed by two sheets of optical glass on whose inner faces have been etched very fine parallel grooves which are filled with black pigment or magenta dye. The distance of the screen from the sensitive film is given by $D = ts/d$ where D is the distance between the black line screen and film, t is the lens extension of the camera, s is the side of the square dot of the screen, and d is the diameter of the lens diaphragm.

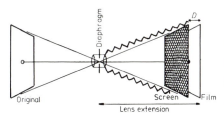

113

(a)

4.7.3

Straight line reversal. The need in the graphic arts field for an optical instrument to give a mirror-reversed picture, as an alternative to the old-style L-shaped camera, is met by a reversal image being produced through three reflecting surfaces arranged mutually at right-angles, the emergent light being parallel to the incident light. The unit as illustrated (a) is in its intermediate position, the upper prism being rotatable about a vertical axis. To obtain reversal, it is turned to face the mirror. The reversing action is illustrated schematically (b). Following the path of the letter R, which is drawn in at various points as it goes through the system, the action should be clearly understood. When unreversed negatives are required, the unit need not be removed. The upper prism is merely turned through 90° to face the copy and the mirror is not brought into use. Illustration (c) demonstrates the alternative light path.

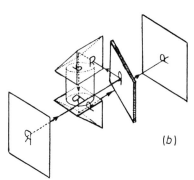

(b)

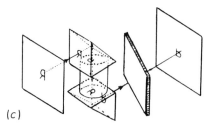

(c)

albumen, fish glue and ammonium dichromate, or resinous compounds known as diazonium salts. During exposure to light, the resist is hardened. The plate is then rolled with a greasy ink and washed to remove the unexposed resist with its ink layer. The plate should then be placed in an oven to further harden the resist and so protect the metal during the etching operation. Zinc is etched with nitric acid, but copper is always etched with a solution of ferric chloride. The resist is then removed from the plate.

As an alternative to the etched plate, various types of photopolymer relief printing plates are available, consisting of a base sheet of aluminium or polyester film, precoated with a layer of polymer. When exposed to ultraviolet light through a negative, the plate image polymerises and becomes insoluble. The non-image area can be dissolved and washed away by a processing solvent of alkaline water or alcohol to leave a relief image, which is then dried and hardened by heat. The minimum highlight dot that can be printed depends on the quality of the negative, the profile of the dots, light scattering within the plate material and processing characteristics.

(a)

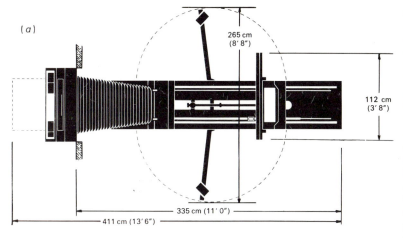

265 cm
(8' 8")

112 cm
(3' 8")

335 cm (11' 0")

411 cm (13' 6")

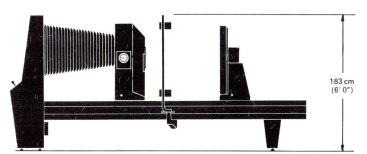

183 cm
(6' 0")

4.7.4

(a) and (b) Lithodex darkroom camera for direct and reverse working. (a) Schematic diagram of the Type 41; the copy holder of the Type 40 is central to the camera body. (b) Type 41 lens panel for direct and reverse working. The vacuum filmholder is for use with contact screen and film up to 50 cm × 65 cm (20 in × 25 in). The standard lens is of 460 mm (18 in) focal length giving 2½× enlargements and 4× reductions. Four 1250 W tungsten–halogen lamps provide illumination. (c) Lithodex 46 camera, made by Pictorial Machinery Ltd, for darkroom operation with film up to 40 cm × 50 cm (16 in × 20 in). The zoned-vacuum filmholder simplifies the use of contact screens, and tungsten–halogen illumination reduces exposure times.

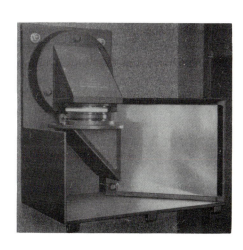

(b)

(c)

115

The papier-mâché sheets are dried centrifugally on the inside of a drum with a curve to suit the casting operation to make a curved stereoplate. The matrices of papier-mâché are then placed in the casting machine where molten metal alloy at over 300 °C is poured on them and solidifies almost immediately. The casting contains the print of the page in negative form, upon the convex side, ready for the plate cylinder of the printing press (Fig. 4.7.6).

4.7.5
Lithodex 26 camera, made by Pictorial Machinery Ltd, will accommodate up to 40 cm × 50 cm (14 in × 18 in) film and copy up to 50 cm × 60 cm (20 in × 24 in).

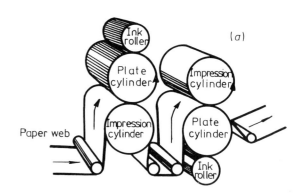

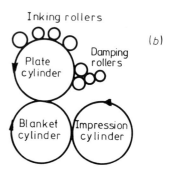

4.7.6
(a) The rotary letterpress printing machine. (b) Offset litho machine. The first cylinder carries the printing plate with its damping and inking mechanisms, the second carries a rubber blanket, and the third carries the paper. Offset lithographic colour-printing machines are usually built in units, each unit printing one colour.

The plate goes to the foundry where it is mounted on a metal slab to the same level as the type produced from line-casting machines and they are fitted together within a steel frame known as a chase. The next stage in the stereotyping process, for making metal duplicates of type, is to place a dampened sheet of papier-mâché on the metal page when, under heat and pressure, an impression is made of the page in positive form.

116

Lithography

Lithography was invented during 1798 by Johann Alois Senefelder (1771–1834) in Munich, and takes its name from the early use of flat Solnholfen limestone slabs as the printing surface. Modern lithography uses zinc or aluminium plates which can be clamped around the cylinder of a rotary press. The principle of this planographic process requires that the image is maintained with a fatty ink whilst the non-image areas are kept free of ink by covering them with a thin film of water. The metal plates are finely roughened in a graining machine to enable them to hold a continuous film of water.

The image on the plate is obtained by contact printing from film, as with half-tone illustrations, but automatic step-and-repeat cameras are used for accurately positioned multiple images of repeat patterns or labels.

In the preparation of a photolithographic plate, no etching is normally required, but the layer of light-hardened diazonium salts or dichromated albumen that remains on the plate after development becomes the printing image when it is covered with a fatty ink.

Half-tone colour-separation negatives are made on panchromatic film through red, green and blue filters and, if a fourth printing in black is required, a yellow filter is used to counterbalance the high blue-sensitivity of the photographic film. A projection camera is needed for enlarging colour transparencies and for the production of screened colour-separation positives and negatives (Fig. 4.7.7).

The pigments used in printing inks must be insoluble in oils, water or other solvents and resistant to fading on exposure to light. The range of colouring agents available for printing inks is less than the range obtainable as dyes for colour photography. Because these dyes can produce more highly saturated colours than those obtainable from printing inks,

4.7.7
(*a*) Littlejohn 'Magnacolor' colour-separation camera for 30·5 cm × 38 cm (12 in × 15 in) transparency in camera head and vacuum baseboard of 80 cm × 100 cm (30 in × 40 in). Colour-separation filters are located in the lens-carriage filter turret. Xenon lamps or tungsten–halogen lamps can be used to illuminate the flat copy originals. (*b*) Close-up of the vacuum baseboard and controls.

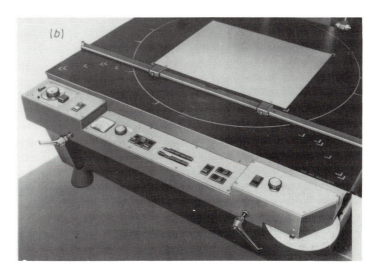

117

a colour transparency may appear more vividly coloured than the same subject printed in a publication. The distortions in tone reproduction, due to the half-tone process, tend to make some greys too dark and others too light, for which corrections are often necessary.

The subtractive primary colours provide yellow, magenta and cyan. Green is reflected from yellow and cyan, red from yellow and magenta, and blue from magenta and cyan. Black is obtainable from the three primary colours and white from the paper makes a total of eight colours which can be produced subtractively by letterpress and lithographic half-tone colour methods. Inks are either fully printed or not at all, so film thickness variations are impossible. Other colours or tones are produced by additive effects, where ink dots are not superimposed, and light absorbance, scattering or surface reflection from the paper influences the colour saturation:

    (a) White minus red is blue-green (cyan).
    (b) White minus green is magenta.
    (c) White minus blue-violet is yellow.

Colour prints on paper use dyes or pigments coloured cyan, magenta and yellow. Colour-reproduction processes are all subtractive but, because in practice colouring materials and dyes cannot be produced with absorptions giving a sharp cut-off curve, it is not possible to reproduce good 'blacks', so four-colour printing methods are used and the fourth ink is black.

In colour printing, the subtractive materials are not mixed, but are overlaid on one another, so they must be transparent and ideally should transmit the light from two-thirds of the spectrum and completely absorb the other one-third of the light.

Coloured materials do not exactly fulfill these requirements, and although the yellow primary colour is reasonably good it absorbs some of the green region. Cyan absorbs too much light and does not transmit enough blue and green. Magenta is reasonably good for red, but does not transmit sufficient blue light.

Because of the unwanted absorption by the magenta in the blue region and of the cyan in blue and green regions, this results in blue, green and purple colours being reproduced, in colour photographs as well as in printing, darker than reality.

The shortcomings of the primary colour materials have to be overcome in photography and printing by various techniques. One method in photography is to use masking by colour couplers in developers, whereas electronic colour scanners offer wide flexibility in the production of high-quality colour printing.

Inks for colour printing were first standardised in Great Britain during 1949 under British Standard B.S. 1480 followed by B.S. 3020 in 1959 and B.S. 4160 in 1967. Materials which are superimposed in a subtractive colour system must be transparent, but the first colour to be printed may be opaque. For this reason, low-cost chrome yellow (lead chromate) is usually printed first, followed by magenta, and then cyan, but transparent yellow inks are available so the colours can be printed in a sequence to suit convenience.

All lithographic printing is produced on offset machines in which the plate is fitted round one cylinder and has to be damped with water as well as inked at every revolution. The inked image is printed on a rubber blanket around a second cylinder and this offset cylinder revolves against the paper carried on a third cylinder and so transfers the ink.

Because the image is offset from a rubber blanket it enables almost any material, as well as all types of paper or cardboard, to be printed in lithography.

4.8 Phototypesetting

The printing of text for books and journals was based on hot-metal case type, for both letterpress and offset lithography, until the introduction of phototypesetting. Letterpress techniques used metal type for the stereotype process, whereas lithography required printed galleys, produced from metal type, for making up into pages to be photographed on film in a process camera.

Composition by the traditional hot-metal type became an obsolescent technology in 1955 when filmsetters were first designed and rapidly developed due to the availability of improved optical systems, zoom lenses for changing focal lengths, and electronic digital computers to control the moving elements.

There were problems in maintaining consistency of image density from phototypesetters, as variables such as exposure, magnification, focus and source of photographic materials affected quality. The design of typefaces had been based on results from metal type and letterpress printing, so changes in the characters were necessary in order to produce good photographic images.

By 1969, computer-aided typesetting, automatic justification, and tabular composition machines had been demonstrated which were capable of phototypesetting speeds up to 40 characters per second. All functional operations of the filmsetter were controlled by a perforated paper tape and data-processing system.

Phototypesetting, with automatic justification, is easy to alter from author's corrections and the exposed photosensitive paper after development is ready for making paste-up pages with or without prints of half-tone screened illustrations.

Phototypesetting machines produce images by means of the following methods:

(a) Contact from a master film matrix.
(b) Projection from a static matrix.
(c) Projection from a moving matrix.
(d) Projection from a moving spot on a cathode ray tube or from a laser beam.

Filmsetters with projection from a static matrix

The Monophoto 400/8 filmsetter has typefaces on matrices for more than 200 designs, including mathematical and foreign languages. An output of more than 40 000 high-quality characters per hour is achieved from stationary images on emulsion-coated film or sensitised bromide paper (Fig. 4.8.1).

Instructions for exposure of 'Monotype' characters are received from a perforated paper tape which selects, by means of an address code, letter columns and number rows on a grid as the basis for the system.

The main advantage of this typographical library is that one matrix can easily be changed for another in the matrix case. The standard matrix case of 400 interchangeable matrices is divided into five areas. Four of these accommodate complete founts of up to 94 characters with the fifth reserved for extra sorts. All founts and sizes can be mixed in a line and from line to line (Fig. 4.8.2).

The filmsetter has a two-part zoom lens system with an iris diaphragm that remains in a correct relative position when point-size changes are made. The light path is directed through an optical flat, then turned 90° by a 45° mirror, passes through the zoom lens to the differential mirror and is reflected by the set-feed mirror to the photosensitive film or paper product. The mirrors have bidirectional movement which allows text to be set in both directions.

The zoom lens provides an automatic character size change from 5 to 24 pt by tape command. Program tapes contain the control data for all typesetting functions including point sizes, line measures, and fixed and justification spaces (Fig. 4.8.3).

The Monophoto 600 filmsetter is a high-quality phototypesetting machine designed to produce all formats of typographic copy from 6 to 28 pt in direct-reading form on film or paper or in reverse-reading form on film. This will depend on whether the output is required for offset lithography or letterpress printing.

Programming and control of the filmsetter is from a 1 in wide eight-channel punched-tape input, generated by a general-purpose computer, or a magnetic-tape reader linked by a connecting cable (Fig. 4.8.4).

The filmsetter carries a fount which is made up of a maximum of 393 standard high-speed matrices and 100 special slide matrices, altogether providing a maximum of 593 characters. The slide matrix area incorporates a drum-type indexing turret having provision for 100 standard 2 in (50·8 mm) square film

119

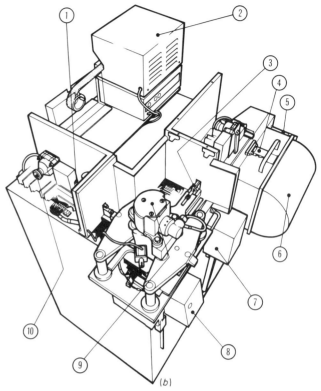

4.8.1
(a) Monophoto 400/8 filmsetter made by the Monotype Corporation Ltd. Exposure takes place through a stationary matrix, and so the exposed image is consistent with the quality of the matrix. (b) The numbered parts on the line diagram are as follows: 1, set-wise positioning ram—Y axis; 2, eight-channel optical reader; 3, point-wise positioning ram—X axis; 4, sheet-film-cassette securing latch; 5, film-feed stepping motor; 6, sheet-film cassette; 7, set-feed stepping motor; 8, point-size change motor; 9, light source and matrix-positioning area; 10, positioning ram dashpot.

4.8.2
Monophoto 400/8 matrix case for up to 400 individual 0·2 in × 0·2 in interchangeable film matrices. The case is divided into four basic founts, each containing up to 94 characters.

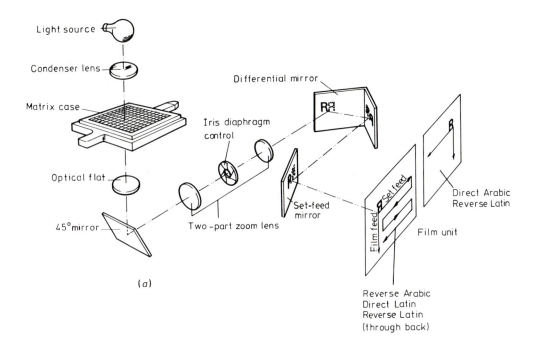

Light source

Condenser lens

Matrix case

Optical flat

45° mirror

Differential mirror

R

Iris diaphragm control

Two-part zoom lens

Set-feed mirror

Film feed

Set feed

Film unit

Direct Arabic
Reverse Latin

Reverse Arabic
Direct Latin
Reverse Latin
(through back)

(a)

4.8.3
(a) Simplified optical system of Monotype
400/8 filmsetter. Interchangeable matrix
cases enable more than 200 typefaces to
be readily available. **(b)** Monophoto
filmsetter in which the characters will be
exposed on film. Alterations to lens and
prism positions change the magnification
of the images. Corrections in filmsetting
are made by cutting and stripping the
errors from the film and superimposing
correct images.

(b)

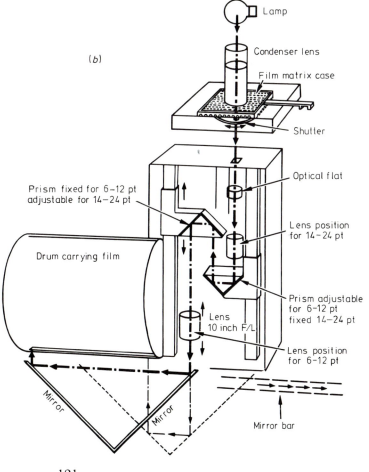

Lamp

Condenser lens

Film matrix case

Shutter

Optical flat

Lens position
for 14–24 pt

Prism fixed for 6–12 pt
adjustable for 14–24 pt

Drum carrying film

Prism adjustable
for 6–12 pt
fixed 14–24 pt

Lens
10 inch F/L

Lens position
for 6–12 pt

Mirror

Mirror

Mirror bar

121

The Monotype 600 filmsetter is programmed and controlled from a 1 in wide eight-channel punched-tape input which has been generated by a general-purpose computer.

slides, each slide accommodating two characters, for composing at about 60 characters per minute. To accommodate this facility, a special square aperture is provided in the No. 4 matrix disc and a primary projection unit is used in association with the flash tube for No. 4 disc (Fig. 4.8.5).

The speed of operation is dependent on the nature of the program, type size and line length being the determining factors, and larger characters or shorter lines will impose lower typesetting speeds. The two film cassettes are similar in appearance except that they are right- and left-handed (Fig. 4.8.6).

The projection system consists of matrix discs which rotate to align the required matrix accurately in the optical paths of flash tubes.

When the required character has been positioned, a short-duration flash from a xenon lamp is projected through it and the resultant image is focused onto photosensitive film or paper. The matrix is stationary when projection takes place, so a well defined image can be achieved. Following each projection, the optical path is advanced along the line of type by a system of moving mirrors, in preparation for the next character exposure (Fig. 4.8.7).

An optical flat, or refraction block, displaces the alignment of an individual character above or below the standard alignment. At 10 pt format, the displacement of a character is possible in the range of 6 pt above and 6 pt below the nominal alignment in steps of $\frac{1}{2}$ pt.

Characters are disposed around the periphery of four matrix discs in an arrangement that ensures that the least call is made for successive characters on the same disc. The characters are selected by the stages of rotation of the stepping motors, one character being allocated to every second step of the motor. Because the inertia created by the masses of the disc, motor spindle and magnets would be undamped, apart from frictional drag in the motor bearings, inertia damping is employed and this uses the inertia of a second mass through a viscous liquid to oppose oscillations of the main mass. This method of damping is adjustable and is applied to all stepping motors.

The characters on the discs are transparent against an opaque background and are illuminated by flash tubes fitted with condenser lenses. The light is collected from the four discs, by a composite glass prism containing semi-transparent mirrors, and passed through a refraction block and telescopic tube to the main projection lens. The light is reflected off two mirrors on the differential carriage and, finally, reflected by the film mirror onto the sensitised surface. A single projection lens is used to obtain different sizes of image from the matrix in the projection system. The simple lens can be represented by a single line, shown vertical within the lens outline, and if the lens focuses parallel rays at a point, then

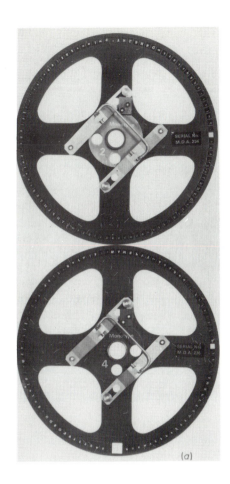

4.8.5
(*a*) 7 in (178 mm) diameter matrix discs. The individual matrices must not be damaged or removed. Each disc has an associated printed circuit board carrying interconnections to allocate unit widths to each matrix position. The square aperture on each disc is a neutral density aperture, whereas the larger square on No. 4 is an aperture for slide matrix projection. Disc No. 2 accommodates 100 individual matrices, whereas No. 4 has 97 only because of the aperture. (*b*) Pi matrix turret with slide ejection slot. Slide matrices carry two characters and encoded character widths on each standard film slide.

4.8.6
Film- or paper-dispensing cassette for filmsetter. Loading must be carried out in a darkroom. Rolls 15 m long and 200 mm wide can be 0·083 mm to 0·127 mm thick. The brake shaft is shown outside of the cassette.

123

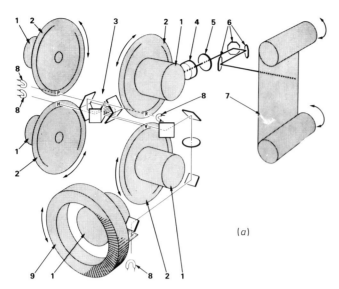

4.8.7

(a) Simplified diagram of projection system on Monotype 600 filmsetter. Each matrix has its own flash tube, one of which is also the light source for the slide matrix turret. The numbered parts are: 1, synchronous motors; 2, high-speed matrix disc; 3, mirror block (simulated); 4, optical flat; 5, focusing lens; 6, set-feed mirrors; 7, product; 8, discharge tubes; 9, pi matrix turret. (b) Four matrix discs and slide matrix drum with matrix finger and ejector.

the distance from the lens to that point is the focal length (f) of the lens. But the film matrix is not at infinity, it is much nearer the lens (distance u) in order to give an image (at distance v) of about the same size (Fig. 4.8.8).

On telescopic equipment, it is possible to adjust a lens combination so that the magnification is varied without loss of focus when, as is the case with the filmsetter, working between fixed object and image planes. Such mechanisms involving lenses are known as zoom lenses and are common on photographic and television cameras in order to produce close-up effects on distant objects. These lenses and mechanisms are complex by nature, in that they involve special cams, cut to match the focal lengths of the lens element, each one requiring individual adjustment. These difficulties have been overcome in the zooming system used in the Monophoto 600

4.8.8
Simplified diagram showing focal length of a simple lens.

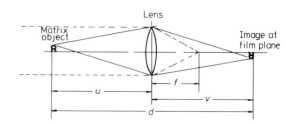

124

by a lens and mirror assembly which has a higher definition than is obtainable by the multi-lens systems and yet is simpler and more compact.

On a simple lens, to change the magnification of an object (matrix) it is necessary to vary the distance (u) from the lens to the object and that (v) from the lens to the image. The sum (d) of the two distances (u and v) is not constant but, by the addition of a pair of movable mirrors, can be fixed with respect to the object plane.

Dimensions w and z are constant and dependent upon general design requirements, so the distance from the lens to the image is now equal to $x + z + y$, and it can be shown that the total distance $x + u$ is a minimum when the magnification is 1 (Fig. 4.8.9). An analogy of the system used can be represented by a lens, locked for a given magnification, to which is fastened a line which passes over a pulley on the differential carriage and thence back to the film mirror (Fig. 4.8.10).

If the lens remains stationary and the line taught, any given movement of the film mirror will produce half that movement at the differential carriage. Consequently, a light beam following the same course will have a constant length. But the lens must move to vary point size and in so doing must activate a mechanism that will produce independent movements of the (differential) mirror pair and the single (film) mirror, in order to maintain focus at all magnifications and while the image conveyed by the film mirror is moved along the line.

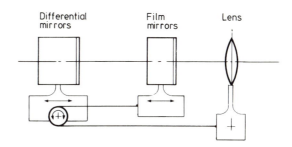

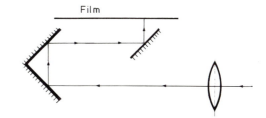

4.8.10
Simplified diagram of differential mirror system.

Laser beam moving-spot character generation

The Monotype System 3000 phototypesetter stores characters in magnetic digital form, not as photographic negatives, and each character of a particular point size is divided into 0·025 mm squares. The top horizontal strip can be recorded as dark or light, then the next strip, and so on down the face of each character. The input information, stored on flexible discs, is read into the main computer store before composition (Fig. 4.8.11).

A low-power helium–neon laser is used for scanning and, because the light is monochromatic at 632·8 nm wavelength, there is no need for colour correction in the optical system. Also, the parallel beam is suitable for high-speed modulation. Panchromatic film or paper must be used because standard phototypesetting materials are not sensitive to red light.

An acousto-optic modulator turns the laser beam on and off following the switching of a signal to the power supply. In the 'on' state, a high-frequency acoustic wave is fed to an acousto–optic crystal which causes variations

4.8.9
The use of mirrors in a lens system.

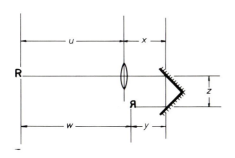

125

4.8.11
Digital representation of a character.

A beam expander for the diffracted light, after emerging from the modulator, produces a wide parallel output beam from confocal mirrors. The input beam must be placed centrally on mirror M_1 in order to fill M_2 symmetrically. The need for a beam expander is a consequence of the wave nature of light, in relation to the high-quality mode requirement, so that a 0·025 mm diameter red spot can be scanned along a 250 mm line (Fig. 4.8.13).

A rotating pyramidal polygon produces the scanning action and each of the eight facets is inclined at 45° to the axis of rotation. Light from the beam expander is incident on the polygon in a direction parallel to its axis of rotation so, as each facet passes through the

in refractive index so that the crystal acts like a phase diffraction grating. The laser beam is diffracted into a number of beams each side of the 'straight through' zero-order beam. By aligning the modulator with respect to the beam at the Bragg angle, most of the light can be made to go into an aperture for the first-order diffracted beam when the oscillator current is on. As the acoustic wave is switched on and off, so the laser light is modulated to expose the moving panchromatic film as a series of small dots (Fig. 4.8.12).

4.8.13
(a) Principle of operation of the beam expander: mirror M_1, radius of curvature R_1, centre of curvature C_1; mirror M_2, radius of curvature R_2, centre of curvature C_2; F is confocal point, i.e. the common focus of M_1 and M_2. It can be shown that $H_2/H_1 = R_2/R_1$. (b) Lasercomp optical system: 1, laser; 2, modulator; 3, filter turret; 4, beam expander; 5, polygon; 6, scanning lens; 7, film moving in direction shown.

4.8.12
The Lasercomp character-generating system builds up the image from a series of scan lines. The scan extends for the full width of the matter being set, so that many characters are being generated simultaneously.

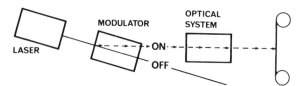

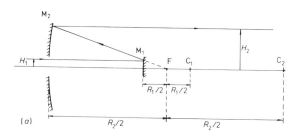

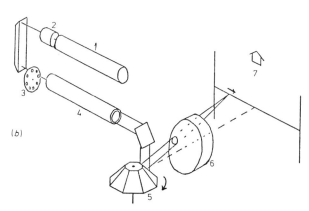

126

beam, the reflected light swings in a plane perpendicular to the axis at an angular rate equal to that of the polygon. Each facet is shaped and aluminised so that the reflected beam is a circle for subsequent imaging to a fine spot by the scanning lens. The scanning lens is of sufficient aperture to collect all of the light as the beam sweeps through its useful angle, and the focused spot scans along a straight line in a flat image plane at a speed of up to 1400 newspaper lines per minute.

Each scan sweeps across the width of the film, perpendicular to its edges, and the film moves lengthwise at such a speed that successive scans produce slightly overlapping exposures. When the first strip is being exposed, the contents of the buffer store are used to determine when to switch the modulator on and off, and the correct position is monitored by a radial grating mounted on the polygon shaft.

Whilst each scan takes place, the next 0·025 mm will have been compiled, transferred to the buffer store, and the film advanced 0·025 mm between the two lines. By a repetition of this process complete lines of text are exposed on the film.

Phototypesetter with projection from a static fount drum

The Mergenthaler variable-input phototypesetters, made by Linotype-Paul Ltd, use an electro-optical system incorporating a film master on a drum which is stationary during exposure and a zoom lens to determine point size of the characters (Fig. 4.8.14).

Film founts, which consist of 96 characters, can be attached to a typeface drum. The number of founts which can be mounted on a drum depends on the point size and range of the characters (Fig. 4.8.15).

The input to the machine is photoelectric from a tape reader to the computer core memory. Justification, from an unjustified tape, can be achieved without a separate computer. Output is printed on 100 ft (30·48 m) long film or paper either 3, 4, 6¾ or 8 in (7·62, 10·16, 17·15 or 20·32 cm) wide.

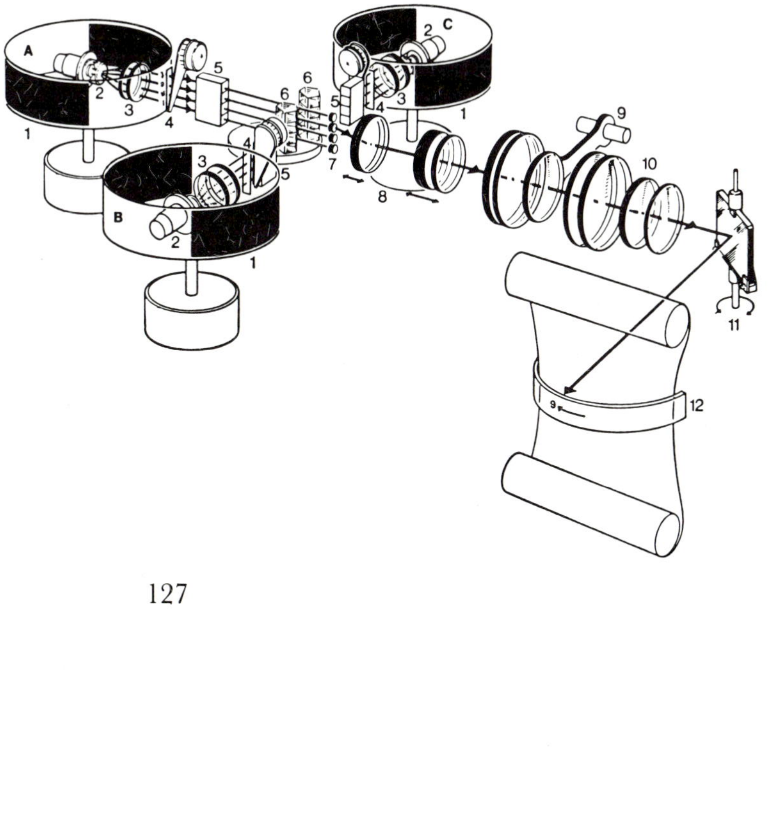

4.8.14
Basic schematic of Mergenthaler V-I-P optical system (three-drum version) showing: 1, font drum; 2, xenon lamp; 3, condenser lens; 4, photocell; 5, aperture plate; 6, pentaprisms; 7, lenslets; 8, solenoid and shutters; 9, magnifier assembly; 10, decollimator lens; 11, rotating mirror; 12, curved film gate.

127

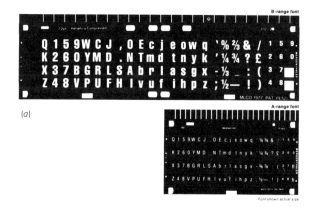

(a)

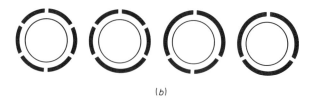

(b)

4.8.15
(a) Typeface designs include Western and Eastern languages and also technical and scientific characters.
(b) Each font drum may be 'dressed' with combinations of A- and B-range fonts: 6 A's, 4 A's and 1 B, 2 A's and 2 B's, and 3 B's, as shown, from left to right.

Cathode ray tube moving-spot character generation

The Linotron 303, made by Linotype-Paul Ltd, is a phototypesetter that contains a character generator which consists of a scanning cathode ray tube, a wheel carrying up to 24 grids and a photomultiplier tube. Electrical 'video' signals from the photomultiplier control a second cathode ray tube which 'prints out' the selected characters on film or photographic paper.

The face of the scanning tube may be regarded as being divided into 144 squares. In any square, the light spot of the tube can draw a series of vertical lines. These are projected onto a grid of 144 characters, and positioned behind the grid is a photomultiplier which converts the vertical lines of the scanning tube from 'light' to electricity. However, the grid of negative characters acts as a 'stencil'—and so the photomultiplier can only 'see' the vertical lines in the image of a character.

The photomultiplier conveys its fine vertical slices of character image to a second, but very much smaller, cathode ray tube. This print-out tube, which is mounted on a moving carriage, sweeps across the required length of line, exposing characters element by element. An optical grating acts as a 'master clock' for the whole system and ensures that the characters are positioned accurately (Fig. 4.8.16).

This phototypesetter will project 137 different point sizes, from 4 to 72 pt in $\frac{1}{2}$ pt steps, and all typefaces or sizes can be used in any

one line. Either 11 or 24 grids can be used on a photo unit and there are 144 characters per grid. Characters for reproduction are high-resolution photographic images. Kodak Phototypesetting film or resin-coated paper, with the developer impregnated in the paper, can be used with this machine.

The Linotron 606 is suitable for large-volume typesetting in full-page composition at a maximum speed of 3000 lines per minute which can include graphics, both line and half-tone, in its full-page output.

Characters for typesetting are held in digital form on a magnetic disc store and there are no photographic grids or matrices. Each fount is digitised for setting at an 'ideal' point size, but may be used throughout a size range from 40 to 175% of the digitisation size. In this way, a fount digitised for 12 pt may be used for a size range between 5 and 21 pt; an original design size of 48 pt provides a size range between 20 and 84 pt, and so on. The number of characters in a typical fount is between 100 and 130, depending on the type style, but the disc storage space occupied by a single fount varies according to the point-size range.

For example, there is room on the disc for approximately

2000 founts for sizes up to 12 pt,
700 founts for sizes up to 24 pt,
250 founts for sizes up to 48 pt, and
125 founts for sizes up to 96 pt.

Expanded and condensed versions of

128

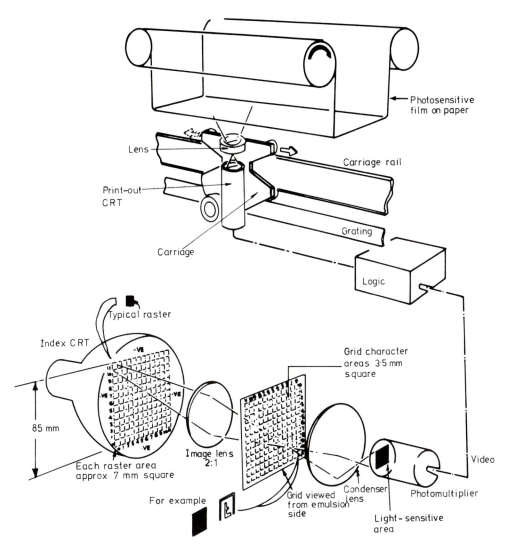

Photosensitive film on paper

Lens

Carriage rail

Print-out CRT

Grating

Carriage

Logic

Typical raster

Index CRT

Grid character areas 3·5 mm square

85 mm

Image lens 2:1

Video

Each raster area approx 7 mm square

Photomultiplier

For example

Grid viewed from emulsion side

Condenser lens

Light-sensitive area

4.8.16
Reproducer optical system.

standard founts may be produced by changing set width independently of point size; thus a narrowing of the normal set width produces a condensed face, and a broadening of the set width makes an expanded face. There are also electronic controls for the production of sloped italic faces from standard founts. Two angles of slope are provided, and the direction of slope may be forwards or backwards.

The basic scanning resolution of the Ferranti Microspot CRT is 650 lines per inch. This resolution is obtained when characters are typeset at the digitisation size. Reduction from the digitised size increases the fineness of resolution and therefore decreases the speed of output in terms of characters per second; enlargement from the digitised size has the opposite effect, giving greater speed but with lower definition.

Users can choose the optimum speed/resolution formula to suit their class of work. In general, speed is the determining factor for newspaper production and fineness of resolution for bookwork.

4.9 Colour scanners

Colour pictures and illustrations on opaque materials, such as paper, cardboard, tinplate or plastics, for magazines, books, journals, catalogues, posters and packaging are reproduced by the subtractive colour process. The word subtractive is used because the pigment or dye subtracts from the illuminating light and we are able to see only the wavelengths which have been reflected. A coloured material selectively absorbs some of the constituent wavelengths and the reflected portions of the spectrum contain the apparent colour of the material.

The apparent colour, known as the hue, usually predominates in a mixture of colours and distinguishes red from orange or green or blue. Yellow is an exception because spectrum yellow is a narrow band of wavelengths and a mixture of red light and green light, of longer and shorter wavelengths, produces the visual sensation of yellow. The effect on an eye from a mixture of primary colours is very important to the technology of colour printing. In 1937, H. E. J. Neugebauer derived mathematical equations for a scheme of colour reproduction, based on eight colour mosaics, and these equations form the basis for colour correction by masking and the use of computers for colour scanning.

In 1948, the first colour scanner was invented in the United States, with colour corrections, based on the equations of Neugebauer, and manufactured by Printing Developments Inc., a subsidiary of Time-Life Corporation. As the electronic digital computer increased in capacity, and complex optical systems with powerful light sources became possible, so the colour scanner was developed.

During 1958, Dr Hell of Kiel produced the Vario Klischograph, which was a letterpress block-engraving device, and one model had colour separation with correction capabilities. This successful machine was also used to produce non-photographic engraved and screened positives which were then contacted onto lithographic emulsions for further processing and production of lithographic plates.

The next stage of development was to make a relatively fast and simple scanner using colour transparencies, and in 1965 K. S. Paul and Associates Ltd produced their first compact colour scanner. In 1967, this private company was acquired by Eltra Corporation of New York, parent of the Mergenthaler Linotype Company. The name of the company was changed to Linotype-Paul Ltd and the colour scanner was marketed as the Linotron 505.

In 1969 the Linoscan 204, the first scanner capable of producing a four-colour set in one single operation, although essentially a same-size machine, was produced. Transparencies could be preplanned and evaluated on spare drums mounted on an illuminated previewing device. Scanning drums were interchangeable in a few seconds. Nineteen electronic masks, including flesh tone modelling, could be preprogrammed for characteristics governed by paper, ink, press and transparency.

Reflection separation copies were made from pigmented, dyed, or screened flexible originals. The colour-splitting prism system eliminated the need for filters and prevented a change of characteristics due to aging. Prefocused tungsten–quartz–halogen scanning lamps and glow modulator exposing tubes ensured rapid service replacement (section 4.11).

The Linoscan 3040 was produced in 1977 to sell at a relatively low price and be simple to operate. Digital processing techniques offered selective colour correction, undercolour removal, and unsharp masking. The computer program provided automatic regulation of highlights and shadows. A size range from 70 to 1000% could be obtained using standard orthochromatic film material with continuous-tone separations on contact-screened positives or negatives. This machine was designed for the medium-sized or small printing establishment where conventional camera processing equipment had become inadequate or overloaded. Technical information concerning this equipment has been provided by Mr Frank Cannings and Mr Tony Waspe of

4.9.1
Linoscan 3040 with transparency scanning on the
left-hand side and exposing of photographic film on the
right.

Linotype-Paul Ltd (Fig. 4.9.1).

The single stabilised xenon light source ensures consistent illumination without variation between scanning and exposing functions. Only one photomultiplier and one electronic channel is involved and the colour separation by prism gives purity of reproduction in terms of colour and tone. Part of the light from the xenon lamp is focused at an aper-

ture to obtain a collimated beam for the scan optics which is reflected by silvered mirrors onto a refracting prism. The spectrum is directed through slots in a disc, spinning at one-third the rate of the scan drum, which has concentric slots spaced at different radii and is positioned so that violet, green and red rays pass through sequentially to a recombining prism (Fig. 4.9.2).

131

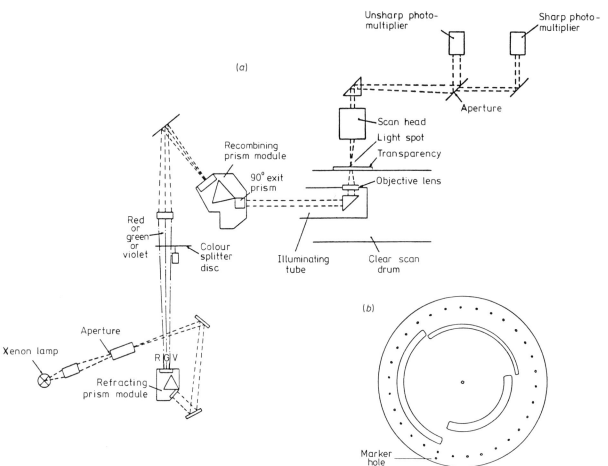

(a)

4.9.2

(a) Scan optics—transparency scanning. The light ray from the optics plate, successively violet, green and red, passes down the scan illuminating tube, is deflected, and is focused by an objective lens onto the emulsion surface of a transparency mounted on the clear scan drum. The scan carriage carries the scan head, sharp and unsharp photomultipliers and a viewing screen.

(b) Disc configuration. The prismatic spectrum is directed through slots in a spinning disc. The slots are concentric to the axis of the disc but spaced at different radii, and also take up approximately one-third of the circumference. The disc is positioned such that the violet, green and red light rays of the spectrum are allowed through sequentially to a recombining prism.

The disc is synchronised with the drum such that only one colour is projected during one revolution, and the changeover from one colour to another occurs at the same point in each revolution. The recombining prism reforms the rays to follow a single light path which is focused to 0·25 mm diameter by an objective lens onto the emulsion surface of a transparency (Fig. 4.9.3).

The scan carriage carries the scan head, sharp and unsharp photomultipliers, and a

viewing screen. Light from the spot on the transparency is collected by the scan head objective lens and focused upon the aperture of an unsharp masking disc. Light surrounding the centre of the scan picture spot is reflected from the white surface of the unsharp masking disc, into the unsharp photomultiplier. Light passing through the centre spot to the reflecting mirror is focused on the sharp photomultiplier tube.

The expose optics system uses part of the

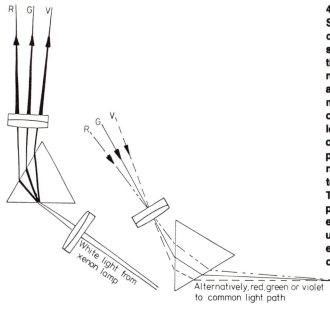

R   G   V

R   G   V

White light from xenon lamp

Alternatively, red, green or violet to common light path

4.9.3

Spectrum ray recombination. The exit lens focuses each colour ray across an object plane, at which point the splitter disc is placed, so that only one colour ray at a time is allowed through in the sequence violet, green and red. At a mid-point between the refracting prism exit lens and the entrance lens element of the recombining prism module, a field lens is placed to commence converging of the previously diverging ray formation. However, the lens is positioned such that the focused but separated colour rays at the disc plane are inside its front focal point. Consequently, onward-projected individual colour rays appear to originate from a virtual image point well to the rear of the plane of the disc, and are divergent. The incident rays at the entrance lens element are projected onto the prism. Since the angle of incidence of each colour ray is different (because they are divergent upon the lens), the amount of refraction experienced by each is also different, and the individual ray paths are combined to follow a single light path.

xenon light which is collimated and passed through a K.D*.P. electro-optic crystal intensity modulator driven by a selected colour signal from the scanner electronics system. A series of expose aperture sizes can be used to suit screen or continuous-tone operation. A circular aperture 0·035 in diameter is used for screen work, whereas a 0·020 in square is used for continuous-tone work. Values of neutral density may be introduced from a circular filter disc, in the optical path, which has 11 filters and one clear aperture around its periphery. The expose drum will accept film sizes up to 12 in × 15 in (30 cm × 40 cm) (Fig 4.9.4).

4.9.4

Expose optics. One of the two fixed-exposure apertures may be pre-selected by a control on the forward face of the carriage, according to whether continuous-tone or screened separations are being made. For screening, a circular aperture is employed, and a square aperture is used for continuous-tone work. Aperture dimensions are such that, at the fixed resolution of 200 lines/cm, line merge is achieved. Values of neutral density filter ranging from 0·1 to 1·5 D may be introduced into the expose light path by means of external control on the forward face of the expose carriage.

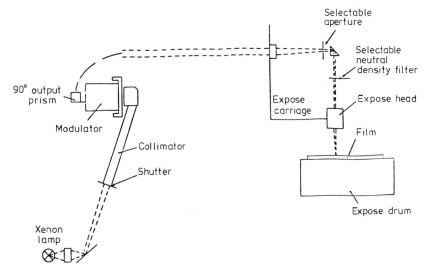

90° output prism

Modulator

Collimator

Shutter

Xenon lamp

Selectable aperture

Selectable neutral density filter

Expose carriage

Expose head

Film

Expose drum

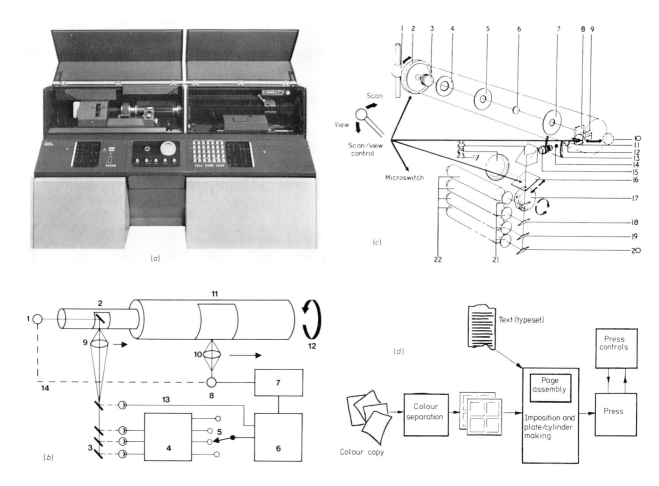

4.9.5

(a) Magnascan 460 enlarging or reducing colour scanner for producing fully corrected screened or continuous-tone positive or negative separations to the desired size in a single step. **(b)** Schematic diagram of the Magnascan: 1, xenon lamp for analysis and exposure; 2, original; 3, colour separation mirrors and photomultipliers; 4, colour computer; 5, four-colour selector switch; 6, tone and UCR computer; 7, digital store for vertical enlargement; 8, light modulator; 9, analysing-lamp advance; 10, exposing-lamp advance for horizontal enlargement; 11, enlarged separation; 12, drum rotation (both drums); 13, unsharp masking

channel; 14, fibre optic light guide. **(c)** Magnascan 460 centre optic unit and analysing unit: 1, xenon lamp; 2, iris diaphragm with uv-absorbing filter; 3, lens 1; 4, light baffle 1; 5, light baffle 2; 6, lens 1A; 7, light baffle; 8, heat-absorbing filter; 9, prism; 10, swinging lens; 11, lens 2; 12, cylindrical lens; 13, transparency; 14, scanning spot; 15, analysing lens; 16, viewing mirror; 17, analysing turret (five positions); 18, red beam splitter; 19, blue beam splitter; 20, plane silvered mirror; 21, correction filters and diffusers; 22, photomultipliers; 23, eye; 24, viewer; 25, prism. **(d)** Colour printing—an overview.

In 1969, Crosfield Electronics Ltd announced the Magnascan enlarging and screening colour scanner. The Magnascan 460 scans any normal original which can be attached to the input drum and produces fully corrected screened or continuous-tone separations to the required size in a single step (Fig. 4.9.5).

Fibre optic light guides conduct light from a high-intensity xenon lamp to the K.D*.P. modulator. A conventional contact screen, as used in a process camera, is superimposed on the unexposed film and vacuum channels in the exposing drum hold the screen and film in contact. Hard-dot positives are produced on the lithographic emulsion film using standard developing techniques.

Scanning speeds up to 4 in min⁻¹ (10 cm min⁻¹) are provided, so A4 sized positives can be produced in 2 min scanning time. The exposing drum will accept film sizes up to 20 in × 24 in (50 cm × 60 cm) and the degree of resolution depends on the enlargement factor, final separation size and output speed. The enlargement ratio is from 0·3× to 16·3×, with almost continuous adjustment over the range of sizes.

The input can be from transparencies or reflection copy originals up to a maximum size of 10 in × 12 in (25 cm × 30 cm), mounted on the analysing drum which can be easily removed from the machine for advance planning (Fig. 4.9.6).

The colour transparency to be scanned and the unexposed film are first mounted on the two rotating drums. Then a small spot of light is projected from the inside of the transparent

4.9.6
The analysing drum on which the colour transparency is mounted to be scanned by a small spot of light projected from the inside of the drum.

analysing drum. The light passes through the transparency into an optical system which splits it into its red, blue and green components, and projects it onto three photomultipliers. The electrical outputs from the photomultipliers are passed to a computer which carries out the functions of colour correction, tone correction, black printer generation, and undercolour removal. The appropriate colour channel is selected by the operator with the colour selector switch. The output from the computer controls the brightness of a spot of light projected onto the unexposed film.

The analysing and exposing drums are coupled together and rotate as one. Each time the drum rotates, one scanned line of the picture is stored electronically and is played back during either the same or the following revolution to the exposing lamp at a different speed, depending upon the enlargement required. The more slowly it is played back, the greater the degree of enlargement. The play-back speed is controlled digitally with extreme precision so that no circumferential distortions are possible. The axial movements of the analysing and exposing optics are provided through lead screws driven by two separate servo motors which rotate at different speeds, depending upon the enlargement required. The speed ratio of the two motors is kept exactly constant by an electronic digital system that provides precise positional accuracy without axial distortion. To change the enlargement, the operator needs only to dial the required ratio into the digital system.

On average, the original pictures are smaller than their final separations. Thus, for economy and speed, the analysing drum is exactly half the size of the exposing drum. By suitably adjusting the play-back and motor ratios, however, it is possible to reduce as well as enlarge. For reflection copy, a light source is mounted outside the scanning drum and the reflected light is collected by the same optical system that is used for transparencies. For unsharp masking, a fourth photomultiplier collects light from the original. The spot of light scanned is slightly larger than the one for the colour analysis photomultipliers and is used to generate a minute fringe around contrasting tones giving the optical illusion of increased sharpness.

Conventional methods of colour correction, such as silver or dye masking, suffer from many well known deficiencies. These deficiencies are usually compensated for by an amount of additional hand retouching. This is particularly so if the original is of poor quality and improvements must be made in colour rendition. Crosfield scanners are fitted with simple-to-use colour controls which may be preset to produce a standard reproduction, or alternatively, be used to provide a degree of electronic retouching that far exceeds the possibilities of photographic methods. Colour controls select the individual pure colours which can be increased or decreased to the required strengths. Each filter separation can be adjusted by six colour controls. In the yellow separation, for instance, the three unwanted colours, i.e. cyan, violet and magenta, can be reduced. Alternatively, these controls may be used to increase the unwanted colour content, if required. This is especially desirable where the processing of the transparency has produced overcorrection. Cyan, for example, can be re-introduced into red areas.

In the same way, yellow, red and green can be increased or, in the case of oversaturated colours, decreased. Combination colours like orange will be automatically corrected. Each colour control has a centre-zero position enabling a standard correction to be set for a given ink and paper combination. If a colour original has abnormal characteristics, however, they can be compensated for by adjusting the corresponding controls to one side of the zero position or the other. After scanning such an original, the colour controls are reset to their centre-zero positions and the machine is returned to standard operation.

In almost every step in the colour reproduction process, some change in tonal gradation takes place. Photographic emulsions and screens are non-linear and the methods of producing printing plates, blocks or cylinders also introduce distortion in the way that printing does. In practice, the worst effect is usually the loss of highlight detail, but there is

also a similar effect on the shadow detail. In most subjects, the shadow detail is less important since it has less visual impact on the eye than the highlights.

Conventional methods of adjusting tone are limited and often difficult to predict. This can cause costly film wastage. Crosfield scanners can be accurately programmed to provide the required tone curves after simple tests have determined the distortions, so that highlight and shadow detail can be retained.

Four individual controls on the scanner provide full adjustment of the tone reproduction curve. Three controls operate in the highlight, middle tone and shadow areas. The fourth control operates in the catchlight-region. In order to simplify the operation of the scanner, when all the tone controls are set to zero, a standard positive or negative reproduction curve is produced that is most suitable for an average original. A further facility provides a different specific tone curve for each colour separation to achieve grey balance. This operates automatically when the relevant separation is being scanned. By changing a plug-in circuit board, the standard reproduction curve can be changed to suit offset, letterpress or rotogravure printing as required. Transparencies which are difficult to reproduce conventionally can frequently be improved by turning the controls from the centre-zero position.

It is generally accepted that the use of a black printer improves the brilliance of print reproduction. Previously, the full effectiveness of a black printer was restricted by the difficulty of producing it photographically or manually. The undercolour removal was also difficult to achieve by conventional means. With an electronic computer, however, a black printer of any required type can be made scientifically and with great accuracy. The corresponding undercolour removal can be carried out with equal ease.

A black printer with undercolour removal has a number of advantages:

(1) It considerably lowers ink costs. In dark print areas, one low-cost black ink replaces three expensive coloured inks.

(2) Difficulties experienced with ink drying on high-speed presses are reduced, thus making possible faster printing speeds.

(3) The registration of fine detail is less critical if it is mainly in the black printer.

(4) A well computed black printer improves the appearance of the reproduction.

(5) It is much easier to maintain the neutral greys during printing.

(6) In wet-on-wet printing, it is possible to limit the maximum ink film thickness to the normal 240%.

The undercolour removal controls are independent of all the other controls and can be adjusted to give 100% UCR. The operator can select any type of black printer from a thin key to full strength black. If required, the black can be removed entirely from the light tones and applied only to the dark tones.

The Crosfield Laserdot electronic screening system can be fitted to all Magnascan 460 machines to improve on the contact screening technique by providing a wide choice of screen rulings, and this will result in enhanced picture sharpness (Fig. 4.9.7).

Laserdot incorporates the well proven conventional screen angles of 15, 45, 75 and 90° to the horizontal for cyan, black, magenta and yellow (parallel to the horizontal) to avoid undesirable Moiré patterns.

The argon-ion laser source with high light output permits the use of normal orthochromatic lithographic film, and the fast exposing time, with elimination of the need to change contact screens between exposures, increases productivity (Fig. 4.9.8).

The Magnascan 550 incorporates digitally controlled and programmed tone and colour calculations which can be applied as needed without resetting or recalibrating the scanner. As all the predetermined tone curves and

(a)

4.9.7

(a) **Laserdot electronic screening system fitted to Magnascan colour scanner. The positive or negative separations are screened at conventional screen angles and correctly positioned for the four colours.** (b) **Laser exposing head. K.D*.P. electro-optic modulators are used to direct the laser beam over the film, in order to expose it, either for continuous tone or screened for a particular colour pattern.**

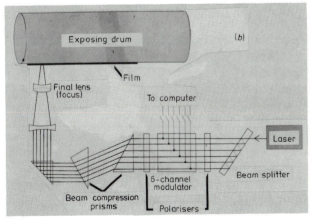

4.9.8
Enlarged half-tone screen. Each dot is generated by the
laser over most of its area, but with a controlled fall-off
around the edges. This provides a compromise between
a hard dot, which cannot be etched without break-up,
and a soft dot, which is etchable but requires critical film
processing.

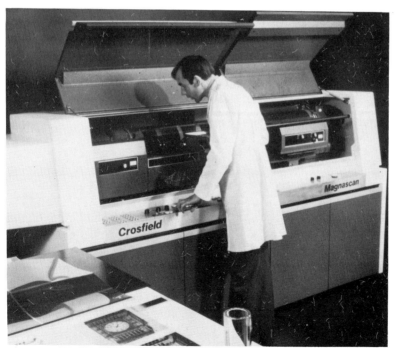

4.9.9
Magnascan 550 colour scanner has
input and output drums of the same
diameter, so four separations can be
produced in one scanning pass
under control of a built-in digital
computer.

138

4.9.10
Input drums. Two sizes of input drums are provided. The small drum accepts transparencies up to 19 cm × 15 cm (7½ in × 6 in) and can be used over the full enlargement range of 0·2× to 19·99×. This is designed for convenient handling when scanning small originals. The large drum, 60 cm × 50 cm (24 in × 20 in), is used for large originals requiring enlargement up to 8·00×. This is particularly advantageous for reflection copy originals or for mounting a number of individual transparencies at the same time. Furthermore, when scanning at same size, four A4 magazine pages may be scanned together.

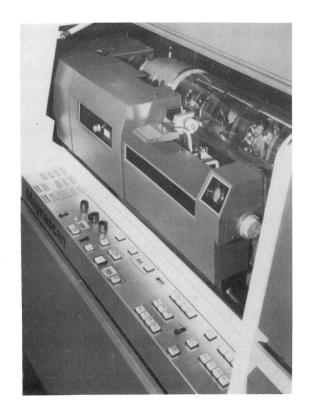

colour information are held in the computer store, if a positive and negative set of separations are required from the same original it is unnecessary to reset the machine (Fig. 4.9.9).

The input optical system has been designed so that all types of film can be scanned satisfactorily, and precautions are taken to avoid bleaching of a transparency left stationary in the scanner (Fig. 4.9.10).

The main feature distinguishing the output side from earlier machines is the provision of a double exposing head on a single drum. Together with the appropriate signal buffer storage, this enables four separations of a size up to one-quarter of the total format to be produced in a single scanning pass. This is known as the QuadraColour system (Fig. 4.9.11).

4.9.11
Output drum. The 60 cm × 50 cm (24 in × 20 in) output drum can be utilised in any one of four different modes: four separations exposed simultaneously to a maximum size of 30 cm × 23 cm (12 in × 9 in); two separations exposed simultaneously to a maximum size of 50 cm × 30 cm (20 in × 12 in); two separations exposed simultaneously to a maximum size of 60 cm × 23 cm (24 in × 10 in); one separation exposed at a time to a maximum size of 60 cm × 50 cm (24 in × 20 in). The diagram shows the exposure of four separations in one pass.

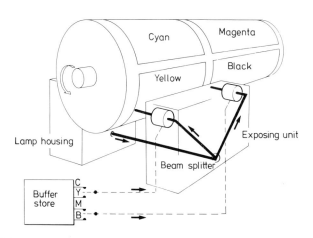

139

Light from a second lamp at the back of the machine passes under the exposing drum through a beam splitter and other optical components which direct it upwards, then horizontally through two K.D*.P. optical modulators and onto the drum. The modulators are independently controlled by the appropriate separation signals switched electronically as the drum rotates.

Thus, the left-hand head normally generates the cyan separation during the first half-revolution, followed by the yellow, while the right-hand head simultaneously generates the magenta and black separations. This arrangement permits a complete set of separations up to 30 cm × 23 cm (12 in × 9 in) to be produced in a single pass (Fig. 4.9.12).

The Magnascan 570 page-composition system, operating with the Magnascan 550, produces a complete set of separations with illustrations and test components in their correct locations ready for copying onto a plate or cylinder.

Originals for reproduction, either transparencies or flexible reflection copy, are mounted on the scanner's analysing cylinder. The operator sets up the tone and colour content of the first original, which is then scanned at high speed and to its final size. Resulting from this operation, picture information at desired final size is stored on magnetic disc and, at the same time, a film separation, which is used only for plotting and not for printing down, is produced. Superimposed on this plotting film is a reference grid structure

4.9.12
The separations exposed on the Magnascan 550 are automatically labelled by colour, and register marks are exposed at the top and bottom of each picture, thus facilitating the subsequent planning operation.

for layout planning purposes in conjunction with the digitising table.

Other originals are then processed similarly, there being no limit to the number of pictures other than as determined by the final page size. According to size, several originals can be scanned onto a single large sheet of plotting film in one operation, thereby minimising film-handling time.

The final-size plotting films are laid on the digitising table in their correct locations for the job in question. A layout plan will normally be present on the table to guide the operator in locating these films. A freely moving cursor is then used to define the relative locations of all pictures and their cropping outlines. In the case of rectangular pictures, the corners are used as coordinates, and placing the cursor on two diagonally opposite corners stores the relevant coordinate data in the disc store alongside the original picture information.

With irregularly shaped pictures, the cursor is drawn around the outline of the picture, in the same way as a retoucher produces a conventional block-out mask, and the coordinate data is lodged in the disc store together with its picture information.

Using the cursor, in association with a simple list of layout commands available on the digitising table surface, the dimensions, colour and positions of tint blocks, rules, borders and backgrounds may also be electronically defined without the need for scanning originals.

After all the layout commands for a page have been entered via the digitising table, there is a short interval during which a page-assembly process occurs automatically in the disc store. In the course of this process, the stored input picture data are assembled, re-configured, and tints, etc, are inserted, in accordance with the stored layout data. Meanwhile, the scanner can be prepared for the subsequent page output scan.

The exposing cylinder of the scanner is loaded with a sheet of unexposed film and, in a single operation, the page information is called from the magnetic disc store and fed to the exposing head of the scanner.

4.10 Lasergravure

Photogravure, or intaglio printing, followed the development of photography and the early history of the process is linked with Henry Fox Talbot (Fig. 4.7.1).

The preparation of an intaglio surface in copper depends on a light-sensitive material, such as dichromated gelatin, being used as an etching resist. A sheet of 'carbon tissue', or 'pigment paper', which is coated with gelatin and coloured to make it visible during processing, is sensitised by immersion in a dichromate solution and exposed under a black-and-white positive transparency. The exposure makes gelatin insoluble to a varying depth according to differing tones on the transparency. A second exposure behind a fine photogravure screen insolubilises the gelatin under narrow transparent lines on the screen and opaque squares form cells on the printing surface.

The carbon tissue is then pressed face down on a copper plate, or printing-machine cylinder, and developed with warm water. The paper backing washes off with the soluble gelatin, leaving a negative relief image of insoluble gelatin on the copper with varying thicknesses depending on the tone values. The network of screen lines are all at the highest level.

By etching the copper with ferric chloride, through the gelatin relief surface, an image is formed in the copper consisting of small cells of varying depth down to about 0·03 mm. The copper cylinder can be faced with chrome plating for very long runs of printing.

In order to be able to use the recessed image at high printing speeds, the ink must be very fluid compared to the thick ink used for letterpress and offset lithography. Because the thin ink must be retained on a rapidly rotating intaglio surface, the whole area must be divided into small cells by means of the photogravure screen. On the press, the cells are flooded with ink and a thin 'doctor' blade, held against the printing cylinder, scrapes surplus ink from the non-image surface.

Gravure is the only process in which tones are reproduced by different thicknesses of ink and this method gives high-quality reproduction of photographic illustrations. The process of making gravure plates has been more expensive than for letterpress or offset lithography, but an entirely new low-cost method of engraving gravure cylinders has been developed using carbon dioxide gas laser technology.

The Crosfield Lasergravure System 700 can engrave a typical cylinder 1·6 m face width and 1·2 m circumference complete in about 33 min. Initially, the copper-plated gravure cylinder is completely etched, so that a conventional gravure cell structure is formed over its whole surface. The cell depths, of approximately 0·05 mm, are not critical and exceed the normal maximum requirements. At this stage, screen angling can be carried out and the cylinders marked accordingly.

The cylinder is then coated with a specially formulated epoxy resin using electrostatic spray techniques, heat-cured and honed to give a non-printing surface. The cured epoxy resin has similar characteristics to the copper base so that, after honing, the surface is flat and without any troughs or crests in the cell structure. Cylinders can be prepared to this stage and stored indefinitely for use later.

The next operation is engraving. During this process the epoxy in the gravure cells is vaporised to the required depth and area by modulating a high-power CO_2 laser beam which is focused on the cylinder surface. The raster of the laser is such that each cell is scanned more than once and the frequency of modulation can provide two 'bites' in the circumferential sense. Each cell can, therefore, receive four bursts of energy and the resulting partial excavation provides for excellent definition of fine-quality work.

During the engraving process, the cylinder rotates at 1000 rpm on a stable lathe bed and the laser optics traverse the cylinder face at a rate of 75 mm min^{-1}.

After engraving, proofing and any necessary corrections can be carried out and the cylinder is then ready for short-run production. For longer runs, when more durability is required, the cylinder is plated with electroless nickel, copper or chromium in the conventional manner. The cylinder now has the outward appearance and lifespan of cylinders made by the traditional process.

After printing, the cylinder is stripped of the plating, refilled with epoxy resin, heat-cured and honed ready for use. As each honing process removes a very small amount of copper, the overall cell structure will eventually have to be re-etched, but it is anticipated that, providing they are not damaged mechanically, cylinders can be recycled up to ten times.

The laser engraving unit can operate from a variety of inputs, thus providing complete flexibility within the system. As the laser is electronically controlled, cylinders can be partially engraved and then completed very quickly, without any masking or staging, when the final pages are available.

The laser engraving machine can be driven in one of two ways. In the simpler system, the engraving unit will be connected on-line to a colour scanner. Where a higher through-put is required, engraving information from the scanner will be recorded on magnetic discs and the engraving unit subsequently driven off-line by a standard magnetic disc pack arrangement.

4.11 Automatic picture transmitting and recording

In a present-day system for picture transmission, the scanning of transmitted copy is invariably by optical means. Typically, the picture for transmission is secured around a drum. A spot of light is focused onto a small area of the picture and, by means of an objective lens, an image of this illuminated area is focused onto a small white screen of about

¼ in diameter. A small hole, or aperture, in the centre of the screen allows the light from a part of the image to fall onto the sensitive area of a photoelectric cell. Any variation in the brightness of the image causes the photocell current to vary, and such variations in current, produced by the scanning of the picture, are used for transmission to the receiver after being suitably manipulated to suit the telephone line or radio circuit which connects the two ends of the system. During transmission, the drum is rotated and the optical system is slowly traversed parallel to the axis of the drum. The entire picture is thus scanned in the form of a close spiral (Fig. 4.11.1).

Receivers of the photographic type, as used by newspapers, are built on similar principles except that the photocell is replaced by a light-emitting device known as a 'glow modulator' tube. In the case of the receiver, the place of the original picture on the scanning drum is taken by a sheet of photographic paper or film.

The rotational speed of the drum and the rate of traverse are chosen to effect a compromise between quality of reproduction and the time taken for transmission. In practice, using normal telephone circuits of 3 kHz bandwidth, a scanning density of 100 lines/in is used with rotational speeds of either 60 or 120 scans/min. These two speeds give transmission times of 16 and 8 min, respectively, for 10 in × 11 in (254 mm × 280 mm) pictures with captions.

To ensure that the received picture is a facsimile not only in its picture content, but also in its physical dimensions, a factor called the 'index of co-operation' is introduced. This factor is derived from the ratio of drum diameter to scanning pitch and controls the aspect ratio or height-to-width ratio of the picture. Provided that two facsimile machines have identical indices of cooperation, then it matters little whether the actual picture size of the received copy is identical with the original as the aspect ratio will be correct and there will be no distortion.

The factors determining drum size, scanning density, index of cooperation and transmission characteristics are regulated by the C.C.I.T.T. (International Telecommunications Consultative Committee) which determines the standards of telecommunication equipment for international usage.

Transmission of newspaper pictures is normally carried on private wire connections using standard audio circuits which are not

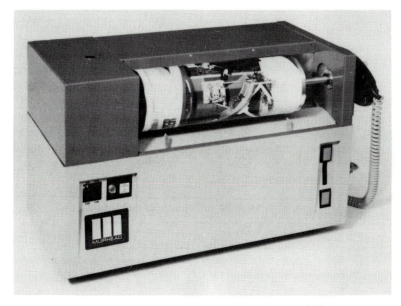

4.11.1
Muirhead K470 high-definition photo-facsimile picture transmitter.

143

usually conditioned to special standards. The public telephone network is not used, except in emergency, since reliable transmission standards cannot be guaranteed on all connections.

There are two basic methods of producing an image for scanning in a facsimile transmitter. One method is to illuminate only the area of the element to be scanned at any instant of time (spotlighting) and the other is to illuminate the general area of the desired element and select the actual element by a configuration of lens and aperture (floodlighting). Floodlighting is the most commonly adopted system since spotlighting requires more accurate alignment, is subject to interference from ambient light, and is wasteful of light energy owing to the difficulty of collecting all the light returning from the illuminated copy.

Earlier forms of floodlight optics suffered from a disadvantage in that, due to the angle between incident and returning light, buckled original copy resulted in false reproduction of photographic tones. Concentric optical systems in which the incident and returning light are on the same optical axis reduce this effect very considerably. The optical unit is sealed and requires no adjustment. It contains the condenser lens, objective lens, mirror and aperture screen. Light from the lamp is collected by the condenser lens and focused onto the drum surface by the front lens. Light from the drum surface returns through the centre of the front lens and the objective lens, and is then turned through 90° onto the aperture screen by a small mirror mounted in front of the condenser lens. The photocell is positioned behind the aperture which determines the scanning spot size of the light received from the drum surface.

At the receiving end, for photographic recorders, the optical system is reversed. The object is to produce a spot of light on the film or paper which is substantially the same size and shape as the original element scanned at the transmitter. At one time, a mirror galvanometer, in conjunction with a system of lenses and mirrors, was used to produce the scanning spot at the receiver. Although many pieces of equipment are still operating regularly

with this system, the crater tube, or glow-modulator, is universally specified for facsimile photoreceivers of present-day design (Fig. 4.11.2).

The crater tube is a form of gas-filled discharge tube. Within the gas-filled envelope, there is mounted a ceramic tube positioned in the centre of the envelope and parallel to its sides. In the centre of the ceramic tube is a cathode, and mounted on the tube at the end nearest the top of the glass envelope is a metal anode cap. On the application of a voltage between cathode and anode, the gas ionises and initiates the cathode discharge which produces a concentrated area of light at the end of the tube. While voltage is applied, the

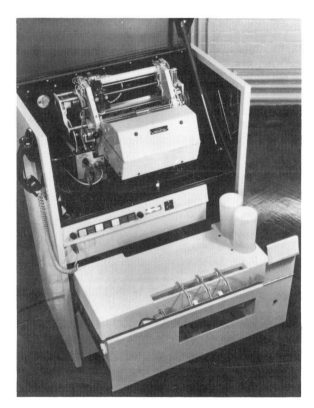

4.11.2
Muirhead K560 automatic picture recorder. The photograph shows the electronics and console in the centre with the processor below at the front. Above the electronics can be seen the cassette for a roll of photographic paper, and above this is the lead screw for the optical carriage. To the left of the screw is the optical carriage itself at the start position.

144

crater tube emits light through a hole in the anode cap, the intensity of the light being proportional to the current flowing. Hence, the light intensity seen through the glass envelope, at the end of the tube, varies in accordance with the transmitted signal. When focused onto the receiver drum, the film is exposed to reproduce a faithful picture of tonal character identical to the original.

Light from the crater tube is collected by the condenser lens, and projected onto the aperture, the intensity of which can be regulated to suit the desired scanning conditions. A final objective lens focuses the aperture onto the surface of the paper or film which is mounted on the receiver drum.

Satisfactory transmission of a picture or chart depends on the transmitted and received material being mounted on drums which are rotated at a constant and identical speed. The modern method of achieving this is to operate the transmitter and receiver by synchronous motors controlled by high-stability tuning forks or crystals. The frequency of the forks or crystals must be sufficiently constant so that, during the time of transmission of the picture, no significant change occurs in the relative phase positions of the two drums. A frequency accuracy approaching 5 parts in a million is required and where supply mains are locked to a standard frequency these can also be used for synchronising.

Phasing can be defined as the adjustment of picture position along the scanning line, or the adjustment of the position of the recording head, to the same location on the recording surface as it is simultaneously held by the scanning head of the transmitter. In practice, a phasing signal is transmitted prior to the commencement of a picture transmission, and is maintained for a period sufficient to enable the receiver drum to be started.

Electronic circuits for facsimile generally follow conventional lines, involving oscillators, amplifiers, trigger circuits and specific requirements such as compensation for the non-linear response of photographic emulsions.

The light output/current input characteristic of the crater tube is substantially linear, and therefore it is necessary to introduce electronic compensation to match the crater tube output to the film characteristic. Adjustable amplifiers with variable gain/amplitude response are employed to suit the overall gain characteristic to the paper or film being used. The resultant compensation can be adjusted to suit the photographic material in use so that correct film exposure results.

Until a few years ago, both transmitters and receivers were manually operated and darkroom facilities were necessary at the receiving end to process the photographic copy which was exposed on the receiver. Today, it is more common for both transmitters and receivers to be automated. Modern receiving equipment uses rapid-access photographic paper developed from the type used for photocopiers. These papers require activation with an alkaline solution and stabilisation in a mildly acid solution to stop the developing process, so can be conveniently processed within the facsimile receiving equipment. Such receivers as the Muirhead K300 series need connection only to a power supply and a telephone line and can be left to themselves to produce pictures sent to them from the distant transmitter. This form of operation is very convenient for newspaper and news agency organisations who operate a worldwide network of news-collecting and distributing service.

The most recent large-scale development in facsimile in the newspaper world is that of remote publication. With circulation increasing in certain specific areas, newspaper proprietors have often found that their distribution means are inadequate. Delivery of large numbers of newspapers by air or surface transport involves delays and expense with a risk of non-publication due to weather or accident. The economics in many cases justify the installation of facsimile equipment which is capable of transmitting whole pages and ensuring faithful copies of extremely high resolution at the distant receiving point. At a receiving point, the facsimile copies, in the form of photographic negatives, are used to produce printing plates from which a complete newspaper edition can be printed. Such facsimile transmissions cannot be made over voice circuits without an excessively long transmission time which could not be tolerated by a newspaper publisher. Wide-band circuits of 48 kHz are therefore used in a multichannel coaxial cable system. In some systems, such as *Wall Street Journal* operation in California, a television channel is utilised. Such high-definition systems require the ultimate in mechanical, optical, and electrical design and are very demanding on the communication circuit between the transmitter and the receiver. The peripheral speed of the scanning drums can approach 100 mph and the scanning resolution required of the system can be as high as 0·0005 in. A transmission system, approaching television in its transmission rate of approximately 2 Mbit s^{-1}, has to be accomplished by mechanical scanning means and a system in which the received image is not a transitory one to be seen and then forgotten but a permanent record of a very high required standard.

Looking into the future when news pictures will be printed in colour, Muirhead have developed facsimile picture equipment which provides, in one single transmission, colour facsimile. A colour print can be faithfully reproduced as a facsimile colour print at any distance and, provided that the line connections from transmitter to receiver are conditioned to C.C.I.T.T. Recommendation M.102 for data transmission, the transmission time is no longer than the normal black-and-white transmission at 60 rpm of the scanning drums.

At the transmitter, the light returned from the copy to the scanning aperture is divided into the three primary colours, red, green and blue. The available frequency spectrum from 300 to 2800 Hz is divided similarly into three channels, each channel carrying its particular colour information. The green channel is the widest of the three and carries the maximum amount of luminance information. At the receiver, the three channels operate three crater lamps, each via their own particular compensated amplifiers, and the crater lamps expose the photographic stock through appropriate colour filters.

4.12 Colour television

Light waves are carriers of information, without the rays themselves being coloured. Colour is subjective and the sensation of colour cannot exist unless seen by an observer. Colour does not exist, even between the retinal receptors and the visual cortex of the eye, except when the information has been interpreted in an observer's brain.

Thomas Young, in 1802, suggested that the retina contained three vibrating systems which responded to the principal colours of the spectrum. However, little notice was taken of Young's suggestion until nearly 50 years later it was put on a formal and quantitative basis by H. von Helmholtz and became the Young–Helmholtz three-components theory. The reproduction of colour by physical and chemical processes in colour photography, colour television, printing, fabrics, paper, plastics and ceramics has become important for both science and technology.

The human eye is incapable of analysing the component wavelengths of the spectrum and is only capable of discerning hue, brightness and saturation. The term 'hue' is that aspect of colour which changes depending on the part of the spectrum radiated and is concerned with the dominant wavelength which is present. Luminous flux causes the sensation of brightness, and saturation is a measure of the purity of the colour which decreases as the colour is mixed with white. The term 'chromaticity' is used to describe both hue and purity.

If three light sources of different hues are mixed, as in colour television, then a whole range of colours can be produced and almost all chromaticities can be simulated. Additive colour mixing can be demonstrated with three projection lanterns and filters, or monochromatic radiations from a spectrometer for red (650 nm), green (530 nm) and blue-violet

(425 nm) wavelengths. The amount of each primary hue to generate a particular colour is known as the tristimulus value.

The International Commission of Illumination (*Commission Internationale de l'Eclairage*) agreed in 1931 to express all colour-mixture data in terms of the three specified component light radiations. White light is defined as the colour of a source emitting a continuous equal spectrum throughout the visible region, and if equal radiant fluxes of the primary red, green and blue-violet lights are additively mixed then white light is generated. In the chromaticity triangle, white light is represented by the point at which the primary light values are all equal to 0·33.

Red, green and blue occupy the three corners and the spectrum is spread over the upper two sides. The magentas or purples on the lower side are called extraspectral colours because there is no single wavelength which by itself will give a magenta sensation, and this can only be produced by a mixture of red and blue light. In additive colour mixing, selected amounts of the fundamental primary colours are added together (Fig. 4.12.1).

Louis Dufay, in 1908, introduced the Dioptochrome plate, which had printed lines of colour, and this was perfected in 1934 as Dufacolour film which was a mosaic-screen additive process. The panchromatic emulsion was exposed through the mosaic and processed by chemical reversal to provide a positive image which could be examined or projected by white light.

The mosaic-screen process of additive colour reproduction is used in television. The colour camera, through a dichroic filter and prism system, separates the picture into primary colours which are detected on three television tubes (Fig. 4.12.2).

The colour-television camera consists of a zoom lens assembly, a colour-separation system, and three colour tubes as described in section 2.5 of *Lens mechanism technology* and section 2.9 of *Dividing, ruling and mask-making* (Fig. 4.12.3).

The three colour separations in the form of electrical signals are transmitted by electromagnetic radio waves in the ultra-high-frequency band and, because of the range of frequency variations in the carrier wave caused by signal modulations, a bandwidth of 8 MHz is needed.

At the television receiver, the picture must be reformed line-by-line by a scanning beam of electrons having energy variations in accordance with the transmitted signal. The three signals actually transmitted consist of two

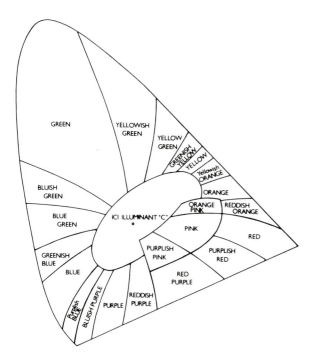

4.12.1
The chromaticity triangle for additive colour mixing.

147

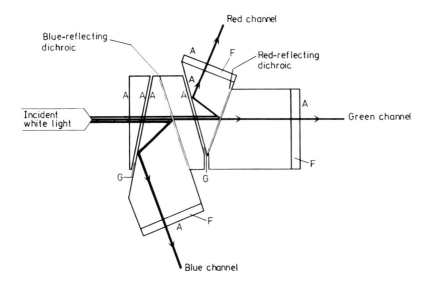

Red channel

Blue-reflecting
dichroic

Red-reflecting
dichroic

Incident
white light

Green channel

Blue channel

4.12.2
Colour separation system.
Colour-splitting prism assembly for
use between television camera lens
and tube as manufactured by Barr
and Stroud Ltd. Faces marked 'A' are
anti-reflection coated; trimming
filters are denoted 'F'; air gaps are
denoted 'G'.

4.12.3
Rear view of prism assembly on television camera
showing the mounting pads for three colour tubes, as
manufactured by Barr and Stroud Ltd.

'colour-difference from white' signals and a luminance signal for image definition information. Because the three primary colours have a constant relationship, it is only necessary to specify two of them in order that the electronic circuits in the receiver can generate the three colour signals.

The cathode ray tube in the receiver has three electron guns which scan phosphors on the rear surface of the screen through a thin steel shadow mask containing about 400 000 holes or slots. Fluorescent powders, consisting of zinc and cadmium sulphides, are termed phosphors when they are mixed to yield colours on television receiver tubes. Typically, the red phosphor may be europium-activated yttrium oxysulphide, the green could be copper-activated zinc-cadmium sulphate, and the blue made from silver-activated zinc sulphide (Fig. 4.12.4).

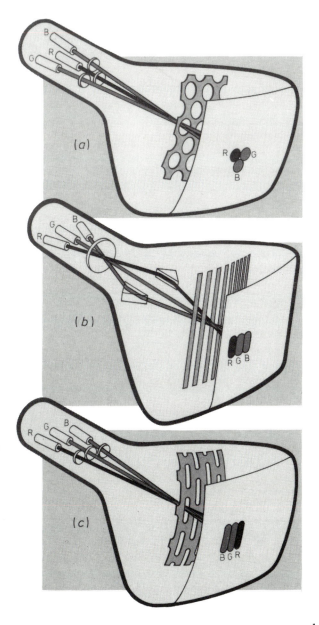

4.12.4
Television receiver aperture masks, as manufactured by the Buckbee–Mears Company, Müllheim, West Germany.
(*a*) Conventional tube. The conventional colour CRT which has been in production since the mid-1950s features a dot aperture mask and three electron guns, which generate the electron beams, arranged in a delta pattern. The glass faceplate has an approximately spherical shape with the mask being mechanically formed to match the contour of the faceplate. The three electron beams are caused to converge at the plane of the mask by focusing electrodes associated with each gun. Control of the convergence angle is necessary to ensure that each beam activates the proper phosphor dot on the inside of the faceplate. TV sets using conventional CRT may require up to 15 convergence control circuits, which must be carefully adjusted on the set assembly line.
(*b*) Trinitron® Tube. This tube, introduced in the late 1960s, features a mask with vertical slits that does not require mechanical forming because of the cylindrical shape of the faceplate. The three electron guns are arranged in a horizontal plane with a unique single focusing electrode system which converges the beams at the centre of the electrode. Since the beams must be converged again at the plane of the aperture mask, a second set of electrodes is used to provide the proper final convergence angle. The phosphor pattern has vertical stripes and requires only two convergence controls in the TV set manufacturing operation. (*c*) Slot Mask Tube. This design was introduced in the mid- to late 1970s. As in the conventional tube, the glass faceplate is approximately spherical in shape, requiring the mask to be mechanically formed. The mask pattern is slotted so as to provide additional metal to impart the strength necessary to allow the flat mask to withstand the forming operation. The plane of the electron guns is horizontal ('in-line') with individual focusing electrodes in each gun. Most tube manufacturers make their phosphor screens in vertical stripes, although the phosphors may also be screened in a 'brick' pattern. The 'slot mask in-line gun' design permits elimination of all convergence controls in the TV set.

®Registered Trade Mark of Sony Corp.

149

5 Projectors for Education and Entertainment

5.1 Introduction

The ability accurately to project developed images is equally important as the initial recording on a photographic plate or film. Projectors are required for 8 mm, 16 mm and 35 mm motion pictures, 35 mm slide projectors, and for use with front or rear projection screens (Fig. 5.1.1).

Rear projectors are used for the inspection of small engineering components, in a workshop or view room, where the enlarged image is compared with a profile or scale on the front of the screen. Measurements on the screen image can be made with precision glass scales, as the divisions on the scale are magnified to correspond with the enlargement given by the lens in use (Fig. 5.1.2).

The portrait lens designed by Joseph M. Petzval in 1840 has been modified for use as the most popular type of projection lens for 8 mm, 16 mm and 35 mm film. A high-aperture lens at reasonable cost is possible with this design, as cine-film projection does not require a wide angle of field. For 8 mm and 16 mm lenses, a field flattener is sometimes added to reduce the Petzval curvature (Fig. 5.1.3).

An entirely different condition applies to the design of enlarging lenses, as there is no need for a large aperture but a very high standard of reproduction is essential. In some cases, it is possible to use high-quality camera lenses under enlarging conditions but, for the best performance, it is usually desirable to construct specially computed lenses.

Camera lenses which have been corrected for distant objects at infinity will not perform satisfactorily if they are used in an enlarger at low magnification values. Enlarging lenses are required for use with printing papers which are sensitive to the blue end of the spectrum, whereas camera lenses have been corrected for panchromatic emulsions. Because of the blue-sensitivity of the printing paper and focusing of the enlarger lens by eye, which is most sensitive to yellow-green, the lens must be achromatised for these two regions (Fig. 5.1.4).

Enlargers, optically similar to projectors, are made for the production of prints from original negatives or microfilm records. Enlargers may have an opal bulb as a light source, and the advantage is that lamp adjustments are not necessary for changes in magnification. The opal electric bulb envelope provides diffusion of the light and prevents the filament showing on the screen (Fig. 5.1.5).

The alternative condenser illuminating system is very similar to that used in a slide projector. An even level of illumination is required over the whole of the negative,

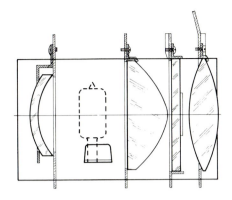
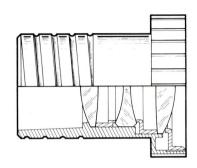

(a)

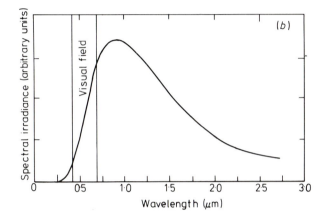

5.1.1
(a) Slide projector optical system including mirror, quartz–iodine lamp, flame-polished aspheric lens, heat filter, condenser lens and plastic-mounted projection lens. (b) Quartz–iodine tungsten-filament source spectral function.

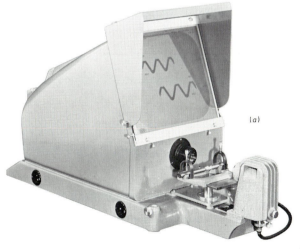

(a)

5.1.2
(a) The Hilger Chekker projector is suitable for checking the dimensions of small components in the workshop and detecting errors before they reach the inspection department. (b) Watts transparent glass projector scale, engraved on the underside, with magnification from 10× to 100× to suit projection lenses.

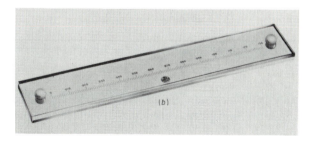

(b)

151

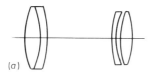

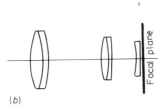

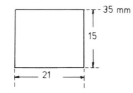

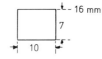

5.1.3
(a) Petzval lens as used for 35 mm motion-picture projection. (b) Petzval lens with field flattener as used for 16 mm motion-picture projection. (c) Motion-picture projector aperture sizes, dimensions in millimetres.

including the corners, so the focal length of the condenser must match the film size and lens aperture to ensure that as much light as possible is directed into the enlarging lens. An adjustment to the light source is necessary for significant changes to the range of enlargement or there will be insufficient illumination at the corners of the frame.

Overhead projectors use a moulded plastics Fresnel condenser system. The Fresnel lens is the equivalent of a conventional lens reproduced as a series of rings impressed on the surface of an acrylic polymer moulding. Spherical, elliptical or parabolic surfaces can be produced with varying degrees of correction to suit the optical system (Fig. 5.1.6).

Fresnel lenses are light in weight and cost about 40% less than the glass lens equivalent in production quantities. A disadvantage is that this type of condenser lens will lose stiffness above 85 °C so it can only be used where there is no excessive heating and the only important use is in overhead projectors (Fig. 5.1.7).

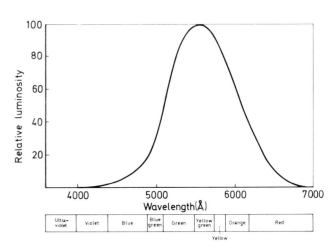

5.1.4
Day-vision colour sensitivity of the eye, with a maximum efficiency at a wavelength of 550 nm.

152

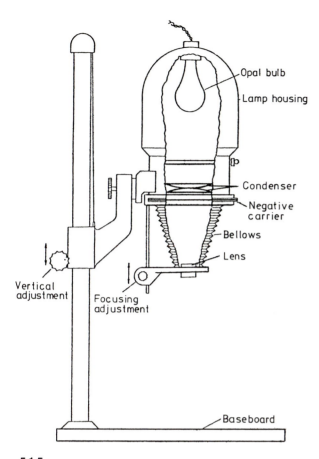

5.1.5
Enlarger

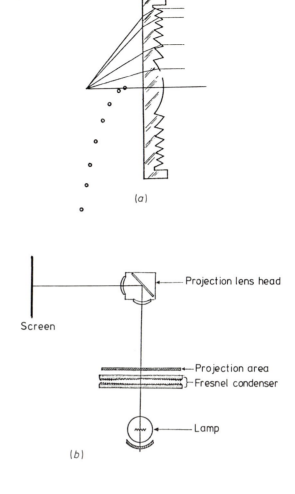

(a)

Screen — Projection lens head

Projection area
Fresnel condenser

Lamp

(b)

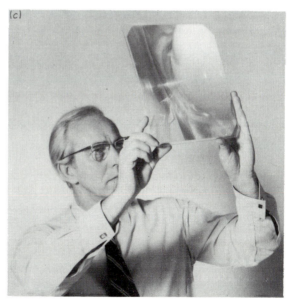

(c)

5.1.6
(a) A Fresnel lens (J. D. Fresnel, 1788–1827) consists of a central spherical or aspherical form with circular zones surrounding the central lens, all having the same focal point and almost the same thickness. The centres of curvature of the circular zones (marked as small circles) do not lie on the optical axis. The radii of curvature increase with distance from the centre as in an aspheric lens. The lens will give parallel rays of light, but dark annular spaces are formed by the dead areas between the circular zones. (b) Use of Fresnel lenses in overhead projector. (c) Pair of Fresnel lenses, cemented around the edges, for overhead projector condenser.

Opal bulb
Lamp housing
Condenser
Negative carrier
Bellows
Lens
Vertical adjustment
Focusing adjustment
Baseboard

153

5.1.7
Rank Aldis overhead projector including a pair of Fresnel condenser lenses, silicon-protected mirror and three-element projection lens.

5.2 Motion-picture projectors

Attendances at cinemas have declined rapidly since 1955, due to the expansion of television, and from 1960 colour feature films have been required suitable for broadcasting (Fig. 5.2.1).

When colour television has for its input material a reproduction of the original scene, by means of a colour motion-picture film, the picture quality reaching the receiver will be inferior to that from the television camera when viewing the scene directly. Television requires very few prints, sometimes only one copy is required, and cannot afford to pay for the reviewing and repeats which are economic when the cost can be spread over a large number of release prints (Fig. 5.2.2).

The high cost of motion-picture photography for colour television is such that 16 mm film is used even in situations where the technical quality of 35 mm film is only just satisfactory. The use of 16 mm film often results in loss of resolution, visibly increases grain, and may cause some picture unsteadiness. From the earliest days of motion pictures, RCA systems have been used in film studios, cinemas and television laboratories. The RCA TP66 16 mm Telecine projector has been designed with built-in automatic features to restore a lost loop in the film and to replace burnt-out projector or sound exciter lamps within one second (Fig. 5.2.3).

The Telecine projector can be amalgamated with a multiplexer, dual-drum 35 mm slide projector and telecine camera TK28 to make an automatic broadcasting system which corrects variations in density and contrast range, flare, film-base errors and improper balance (Fig. 5.2.4).

The three-tube Telecine colour camera has three colour-correction circuits which make it possible to match the colorimetry of other cameras, compensate for variations in colour-film stock, correct for film with low colour saturation or adjust improper printing. The saturation and hue of the three primary colours, red, green and blue, and their complementary colours, yellow, cyan and magenta, are corrected without affecting the luminance grey-scale signal.

154

5.2.1
Pairs of projectors for 35 mm film and 16 mm film in a large cinema installed during the period 1953–58. Carbon arc lamps were used for illumination, and controls were all operated manually. In the late 1950s, xenon arc lamps replaced carbon arcs and automatic changeover controls were introduced.

5.2.2
35 mm optical sound film projector fitted with endless looping attachment for preview theatres where advertising films or short lengths of feature film are required for display.

(a)

(b)

(c)

5.2.3
(a) RCA Telecine projector TP66 for 16 mm optical or
magnetic sound film with capacity for 100 min running
on one reel. (b) Automatic projection-lamp change
mechanism places new lamp into position within one
second of lamp failure. (c) Automatic sound-exciter-lamp
change mechanism on 16 mm optical soundhead.

Motion-picture projectors, with optical heads for sound reproduction in cinemas, depend for their quality on the following factors:

(a) The quality of the sound track.
(b) The sound scanning system.
(c) The properties of the amplifiers and loudspeakers.
(d) The acoustical properties of the auditorium.

At least 90% of the films shown in cinemas have only one optical sound-track. Optical prints often give a higher sound distortion than magnetic prints, and this is usually due to halation of the optical negative, which can be compensated by careful choice of negative and positive densities (Fig. 4.3.1).

With variable-area sound, there will be cross-modulation, and, with variable-density sound, intermodulation, depending on the care taken during printing at the film processing laboratory. The requirement of both optical and magnetic sound is that the film speed, at the point of scanning, should be constant. The term 'wow' is used to refer to undesired deviations of speed occurring at frequencies below 20 Hz and 'flutter' describes deviations of speed in the range 20–200 Hz. The magnitudes of the deviations of speed are usually expressed as percentages of the mean film speed (Fig. 5.2.5).

Slow speed variations, resulting in wow, can be avoided by fitting a heavy flywheel to the film sound drumshaft. The elimination of flutter is not so easy, as it is caused by variations in speed caused when the teeth of the sprocket engages film perforations in order to pull the film through the soundhead. As the film is usually taut when running through the soundhead, the shocks caused by the teeth engaging perforations propagate along the film to the scanning position and so cause flutter.

In order to reduce flutter, a tension roller, which absorbs the shocks, is fitted between the sprocket and the scanning site. This roller must be damped in order to avoid decaying oscillations which will be detected by the soundhead. If the design is to make use of a

5.2.4
RCA Telecine colour camera with automatic colour balancing by continuous sampling.

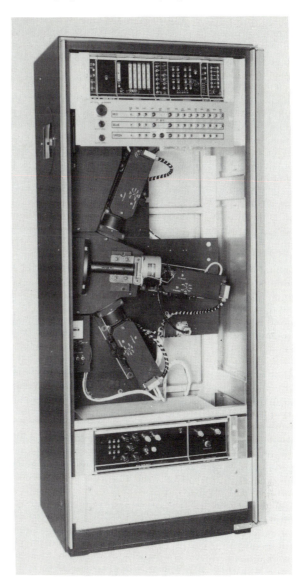

slack piece of film as the resilient element, the correct choice of scanning site and shape of film loop is of utmost importance. The film, by its natural curvature, should lie against the sound drum (Fig. 5.2.6).

5.2.5
Optical soundhead where the film speed must be constant at the scanning position.

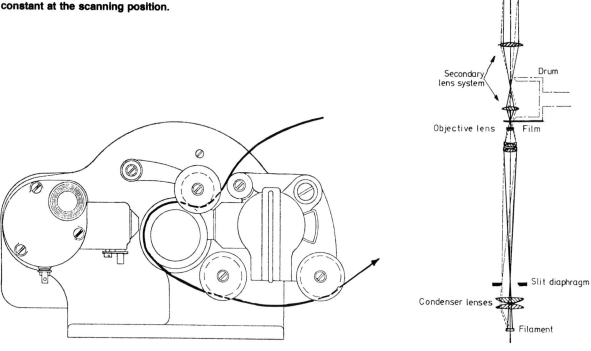

The frequency response of the optical soundhead, which depends on the length of the slit, falls off rapidly at frequencies over 6000 Hz with a 30 μm wide scanning slit and photomissive caesium-coated cathode vacuum cell. With modern solar cells, which have little background noise and do not need a supply voltage, good reproduction of frequencies up to 10 000 Hz is possible with a 15 μm slit length.

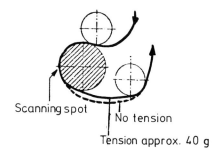

5.2.6
In order to reduce flutter, a tension roller is inserted between the sprocket and the scanning site, and is correctly positioned so that the natural curvature of the film presses it against the sound drum.

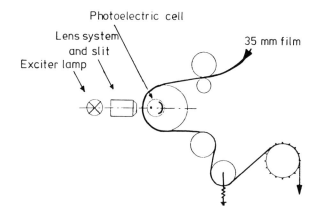

5.3 Printing-press register controls and web viewers

The printing of four colours in sequence on a rotary press depends on the accurate registration of metal plates mounted on the rotating cylinders. Exceptionally accurate control is possible with mark recognition techniques for multimark operation, and a simple push-button line-up method for single-register-mark printing (Fig. 5.3.1).

The Crosfield Autotron 150 is an integrated-circuit automatic register control for web-fed presses. The detection of register marks on the web is performed by a photo-electric scanning head. The Type 173 universal head will register marks across the web or in-line ahead, with magnetic switch or encoder gating alternatively in-line ahead with mark-recognition gating (Fig. 5.3.2).

A five-cell head is used for register control with marks across the web, using mark-recognition gating, and the head contains a recognition system and a register system. The register system uses two photoelectric cells and each cell receives light which has been diffusely reflected from the associated light line on the web. The recognition system contains a five-unit photoelectric cell and light is diffusely reflected by the web from a 29 mm light line through three pairs of spherical lenses (Fig. 5.3.3).

Each outer cell receives an image of a 3·2 mm section of the 29 mm line through a corresponding pair of lenses. The central three cells receive an image of a 9·5 mm section of the 29 mm line through the centre pair of lenses.

The mark-recognition across the web scanning head is laterally positioned so that the two register marks pass under the two register light lines and so one of the register marks passes under the centre pair of spherical lenses. This register mark is recognised each time it passes through the 29 mm light line, just before the two register marks pass under the light lines, and a sequence of events takes place:

(a) All five cells see white paper.
(b) Three centre cells are simultaneously darkened by the register mark, for a distance equal to the mark thickness, whilst the two outer cells continue to see white paper.
(c) All five cells see white paper.
(d) Recognition is completed and a signal is passed to the register circuits just before the register marks pass under the light lines.

For packaging printing, register marks are usually laid down in each colour, and

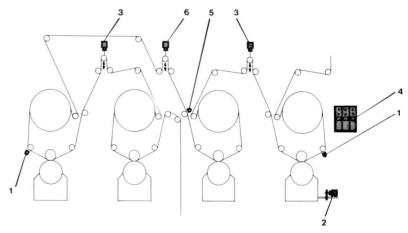

5.3.1
Autotron 150 controlling colour length and back-up register on a rotogravure press. *Colour register control:* 1, scanning heads; 2, digitiser; 3, register motors; 4, electronics cubicle with control panel. *Back-up register control:* 5, scanning head; 6, register motor.

159

5.3.2
Type 173 (universal) scanning head.

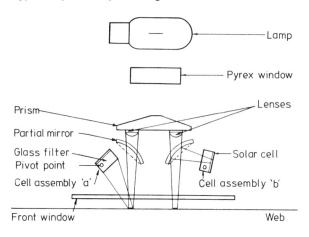

positioned one in front of the other in the margin at the edge of the web. The marks need only to be 10 mm × 0·75 mm with a space of 20 mm between leading edges. Two-channel scanning heads are mounted on the second printing unit and each of the units that follow. These heads are of a new universal design incorporating two photocells which can be set in either one of two positions, thus providing optimum signals when used with

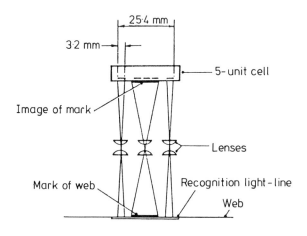

5.3.3
Recognition using a five-unit photoelectric cell.

the very wide range of base materials and inks employed in the packaging industry.

The electronic mark-recognition circuits eliminate the need for magnetic switches or similar gating devices. For single-register-mark operation (including register of varnish and lacquer), digitiser units provide signals corresponding to subsequent printing cylinder positions. These signals are compared with scanning head pulses to maintain register.

The Crosfield Synchroscope 1800 is an optical web viewer that allows the printer to inspect every detail on a fast-moving web for register and colour accuracy. A colour-comparator copy holder can be fitted, illuminated by the web lighting unit, which enables the operator to compare the colour on a moving web with that of the pass copy (Fig. 5.3.4).

The Synchroscope consists basically of 15 strip mirrors mounted around a drum driven by a selsyn motor which rotates at one-fifteenth of the speed of the printing cylinder. Each mirror in turn reflects the same portion of the web, and produces a stationary image onto a fixed mirror which is seen by the printer.

Since the image is continuously visible, the flickering, which renders electronic stroboscopes unsuitable for web viewing, is almost eliminated. A similar effect can be produced more cheaply using a single oscillating mirror and a fixed mirror, but the vibration of a single mirror reduces sharpness of the image and increases wear at the higher printing speeds.

The drive consists of a selsyn generator which can be driven from any convenient point on the press, and is electrically connected to a selsyn motor which drives the mirror drum. In this way, complicated mechanical couplings are eliminated.

The Synchroscope is electrically coupled to the press and can give a stationary image at any press speed. A handwheel is provided to enable the operator to select the strip of web he wishes to view. When the Synchroscope is switched to 'Automatic Drift', a small electric motor slowly turns the handwheel so that the whole printed pattern is continuously displayed.

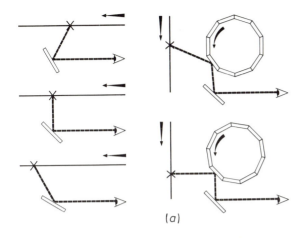

(a)

(b)

5.3.4

(a) This is the basic principle used in the Synchroscope. The diagram on the left shows, in simplified form, how it is possible to stop motion visually. It will be noted that the eye has not moved and yet, by slight rotation of the mirror, it is able to see continuously point X while it is progressing. On the right, it is shown how this principle

is applied in a practical manner. Successive mirrors mounted on a drum and rotated in synchronism with the speed of the web create a stationary picture of the printed material. (b) Crosfield Synchroscope optical web viewer which allows a printer to inspect every detail on a fast-moving web.

5.4 Planetaria

V.E.B. Carl Zeiss Jena manufactured their first planetarium in 1923 for the Deutches Museum in Munich, to the design of Walter Bauersfeld (1879–1959), and since then more than 300 projectors have been delivered throughout the world. The planetarium is used for the education and entertainment of those who are interested in scientific astronomy, or maritime and space navigation.

For instruction in mathematical astronomy, where usually only the fixed-star firmament together with the Sun, Moon and planets are required, a medium-sized planetarium is sufficient. This instrument has one fixed-star sphere only but, as with larger machines, the firmament is produced by the projection of star plates with the aid of high-quality optics.

The northern and southern fixed-star sky is projected from a 220 mm diameter sphere which rests on a structure positioned in the centre of the room. On the surface of the sphere, there are 31 projectors, equipped with

lens systems of 30 mm focal length each, and these represent about 5000 stars down to the sixth magnitude in their correct relative positions and brightness ratio (Fig. 5.4.1).

The Spacemaster projector was designed for installation in an existing building with a suitable interior. The instrument is mounted on four axes:

(a) The polar axis for demonstration of the diurnal motion of the sky.

(b) The ecliptic axis for motions of the Sun, the Moon, the planets and the fixed stars.

(c) The horizontal axis for changing the geographical latitude.

(d) The vertical axis for a horizontal rotation of the entire sky.

The four axes not only permit reproduction of the apparent geocentric motions, as seen at any time and from any place, but also such motions and observational conditions that have been or will be experienced during

161

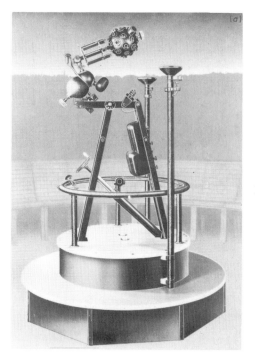

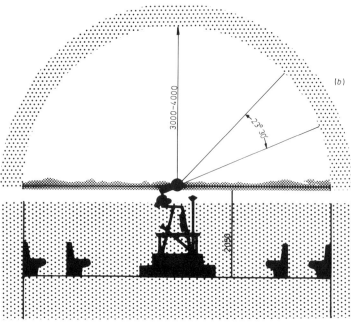

5.4.1
(a) Medium-type planetarium for a projection dome of 6
to 8 m diameter to accommodate an audience of not
more than 80 people. (b) Diagram demonstrating the
design of a planetarium dome.

spaceflights. For example, it is possible to display the motion of the sky as observed from a satellite, from the Moon, or from any planet. The projector can be manually operated or automatically programmed with a tape-recorded lecture presentation (Fig. 5.4.2).

A Universal Large Planetarium requires a specially erected building, in a densely populated city, where the high investment can be made to pay with audiences of up to 600 persons. A structure rotating about the vertical axis holds the principal unit for the different projectors that have to be moved. The horizontal axis for the polar altitude is seated in this structure 3 m above the flooring. Arranged about the horizontal axis is the middle part, which contains the gears and

motors for the diurnal, annual, precessional and polar altitude motion, and the projectors for the various astronomical supplementaries. Attached to one side of the middle part are structures with the gears and projectors for Moon, Sun and Saturn, and on the opposite side are the gears and projectors for Mercury, Venus, Mars and Jupiter. Adjacent to the structures is one sphere carrying the projectors for the northern and another for the southern fixed-star sky. At the extreme ends of the 4 m projector unit are small spheres with projectors for the constellation figures. The lower part of the instrument, which carries the Planetarium projector, is provided with a fairing, housing 28 devices for illumination of the auditorium and dome (Fig. 5.4.3).

162

5.4.2

(a) Spacemaster planetarium projector for domes of 10, 12·5 or 15 m diameter under which 300 persons can be seated.
(b) *Main projecting instrument* (gearings): 1, gearing for Saturn motion; 2, orbit of Saturn round the Earth; 3, driving shaft for planet structure (north); 4, gearing for motion of the Sun; 5, gearing for motion of the Moon; 6, gearing for regression of the lunar nodes; 7, driving gears for planet structure (north); 8, intermediate gear wheel for annual motion; 9, pair of bevel gears for drive of the date projector; 10, bevel gear for annual motion; 11, planetary gear; 12, electromagnetic coupling for drive of the mean Sun and the hour circle; 13, gear drive for mean Sun and hour circle; 14, driving motor for annual motion, with reducing worm gear; 15, connecting gear of diurnal motion to the annual motion; 16, gear drive for diurnal motion; 17, worm gear for precessional motion; 18, driving gears for planet structure (south); 19, gearing for Mercury motion; 20, driving shaft for planet structure (south); 21, gearing for orbit of Mercury and Venus round the Earth; 22, gearing for Venus motion; 23, gearing for Mars motion; 24, gearing for orbit of Mars and Jupiter round the Earth; 25, gearing for Jupiter motion; 26, planet structure (south); 27, Jupiter projector; 28, Mars projector; 29, Venus projector; 30, Mercury projector; 31, connection axle between northern and southern planet structures; 32, driving motor with worm gear for precessional motion; 33, driving motor with worm gear for diurnal motion; 34, driving motor with worm gear for latitude motion; 35, worm gear for latitude motion; 36, gear drive for pole marking and hour–angle scale; 37, gear drive for transmitter of polar altitude indication and map projector; 38, gear drive for date projector; 39, Moon projector; 40, Sun projector; 41, Saturn projector; 42, planet structure (north).

(a)

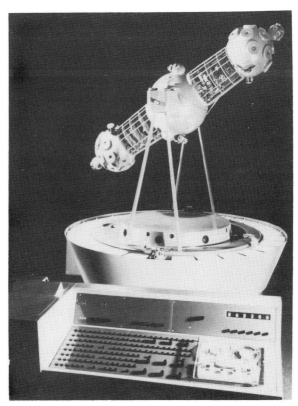

(b)

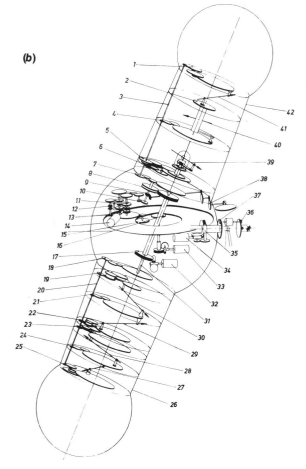

163

5.5 Light sources

Tungsten-filament lamps

Tungsten–halogen lamps are a special class of incandescent filament lamp with bromine or iodine compounds added to the usual filling gases. Quartz–iodine tungsten light sources give at least double the life of a conventional lamp of similar colour temperature. When a tungsten filament is operating at a high temperature, it slowly evaporates and particles of tungsten condense on the bulb wall to cause blackening. With the addition of a halogen to the gas filling, the tungsten particles combine with it to form a tungsten halide. In constructing the lamp in such a way that the wall bulb temperature is kept above 250 °C, the tungsten halide is prevented from condensing and is carried back to the vicinity of the filament. The high temperature of the filament causes the tungsten halide to break down, into tungsten and the halogen, with the result that tungsten is deposited on the filament and the halogen is released to repeat the cycle. Tungsten–halogen lamps do not blacken but emit their full light output throughout the whole of their life. The quartz–iodine tungsten light source is a substandard blackbody at 3000 °K which does not give visible light at the peak of its spectral distribution, so most of the power fed to the lamp is used to produce heat radiation and not visible light (Fig. 5.1.1).

Low-pressure gas discharge tubes

Spontaneous monochromatic light sources, obtained from electric discharges in gases or vapours, are frequently used in scientific instruments as they have strong dominant spectral lines instead of the continuous spectrum of white light. When combined with a variable entrance slit, the light can be resolved into monochromatic lines or bands depending on the circumstances. The dominant 'D' lines of sodium lamps (yellow) at 589 and 589·59 nm wavelength are known as doublets and average 589·3 nm, whilst low-pressure mercury lamps contain strong visible lines at 576·96 and 579·06 nm (yellow), 546·07 nm (green), 435·84 nm (blue-violet) and 404·66 nm (violet). Helium lamps are illuminated at 587·6 nm wavelength.

Fluorescent tubes

Fluorescent tubes are electric discharge tubes in which the internal surface is coated with a substance which becomes fluorescent and emits visible light when it is excited by ultraviolet radiation. Their spectral emission is a mixture of continuous and band emission in the visible wavelength (see section 3.6) (Figs. 4.11.2, 8.8.2 and 8.8.3).

High-pressure gas discharge tubes

Mercury high-pressure gas discharge lamps provide very intense sources of near-ultraviolet and visible radiation as a line spectrum superimposed on a continuous spectrum. The spherical main portion of the bulb contains a basic gas at less than one atmosphere pressure, but when the lamp is energised an ionised column of mercury vapour, with high current density, is established between the two electrodes. This increases the vapour pressure at an operating level to about 40 atm yielding high luminous efficiency. The spectral lines are broad, and the continuous background is bright, but individual lines at 365·0 nm (ultraviolet), 404·6 or 435·8 nm (blue), 546·1 nm (green) and 578·0 nm (yellow) can be selected by glass or dyed gelatin filters (Figs. 5.5.1 and 8.8.5).

Xenon gas lamps

Xenon arc lamps combine high brightness and point source features of the mercury arc lamp, with a continuous type of spectral output approaching that of natural sunlight at an equivalent colour temperature of about 6000 °K. Starting is almost instantaneous and brightness remains essentially constant during the useful life of the lamp. The xenon lamp is cold-filled with several atmospheres pressure of xenon gas which increases to about 30 atm at operating temperature. The xenon lamp can be modulated by electronic means to frequencies up to 30 kHz with lamp light output proportional to input power (Figs. 4.6.13, 4.8.7 and 4.9.1).

Gas lasers

Gas lasers provide a stable high-energy monochromatic continuous source of coherent light and the most important employ carbon dioxide at 10·6 μm, helium–neon at 0·6328 μm and argon–ion at 0·4880 and 0·5145 μm wavelength (Fig. 4.8.12).

Carbon dioxide lasers with gallium arsenide modulators are used in Lasergravure processing machines and argon-ion lasers with

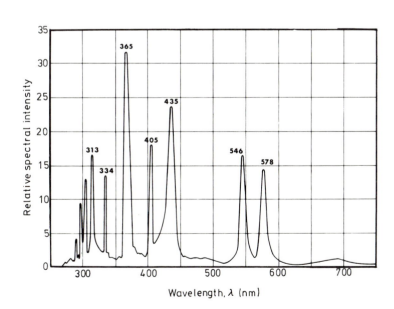

5.5.1
Spectral emission from mercury high-pressure gas discharge lamp HBO 200.

165

K.D*.P. modulators control scanners for making colour separations. Helium–neon lasers are used in phototypesetters for exposing panchromatic film and also for distance measurement.

In a gas laser, there is a mixture of active lasing material, or 'dopant', and a host material contained in an optical cavity. An external electrical source is used to stimulate the lasing process. In this process, dopant atoms absorb energy and, when an atom in an excited state is struck by another energy particle, the outgoing light wave has the energy of both the incoming and absorbed particles. The light wave travels back and forth between mirrors, at each end of the lasing cavity, and when the majority of atoms reach an excited state the laser light is released from the partially mirrored end of the cavity (Fig. 5.5.2).

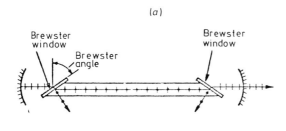

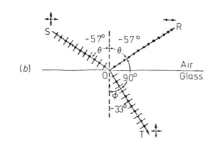

5.5.2

(a) Gas laser showing how the gas tube is terminated with Brewster windows. (b) Brewster's law shows that the angle of incidence for maximum polarisation depends only on the refractive index n as

sin θ /sin $\varnothing$ = n. Since at θ = 57°, $\varnothing$ = 33°, the angle ROT = 90°, we have sin $\varnothing$ = cos θ, giving

$$\frac{\sin \theta}{\sin \varnothing} = \frac{\sin \theta}{\cos \theta} = |n| = \tan \theta.$$

Semiconductor photoemitters

Gallium arsenide electroluminescent diodes emit radiation at particular wavelengths as a direct conversion of electrical energy. This radiation depends on the chemical composition of the host crystal and the n-type dopant atoms of silicon, germanium, selenium or tellurium, which are included as an impurity. Semiconductor diodes are used as incoherent emitters or as lasers when each end of the junction must be cleaved in order to form a cavity where the lasing process can develop. The region of the p–n junction where laser action takes place is only about 1 mm across and 1 μm wide (Fig. 5.5.3).

Gallium arsenide electroluminescent diodes at the infrared wavelengths of 885 and 910 nm are used in rangefinders. High power outputs are possible with an efficiency approaching 50%, compared with 5% for helium–neon gas laser, and modulation of the light output is possible. The high-radiance light-emitting diode has an advantage over the injection laser in that it is capable of both analogue and digital modulation (Fig. 6.4.7).

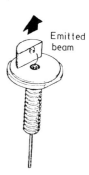

5.5.3

A gallium arsenide injection laser. Tellurium is added to replace some of the arsenic atoms and create an n-type region, whereas zinc is added to replace some of the gallium atoms to form a p-type region. Typical dimensions are 0·1 mm × 0·1 mm × 1·0 mm in the final cut-and-polished crystal. The initial radiation, which is amplified by lasing, comes from spontaneous non-coherent emissions similar to an emitter. The amplification is more than 1000 times and occurs only at the junction area where carriers are able to recombine between the faces of a Fabry–Pérot cavity formed by two parallel polished sides of the semiconductor chip. The amplified radiation is concentrated in a small area and leaves the cavity through the partially transmitting side of the chip.

6 Surveying and Levelling

6.1 Introduction ━━━━━━━━━━━━━━━━━━━━━━━━━━━━━━━

From the earliest days of man, there has been a need to construct dwellings and measure distances with some understanding of the truly vertical and horizontal. As buildings became more complicated, it was necessary to have standards of length and develop geometry for the measurement of angles. The construction of the pyramids by the Ancient Egyptians is clear evidence of the knowledge available to surveyors more than 4500 years ago.

Great skill was used, with the simplest of instruments, to construct large and complicated buildings such as castles, temples and abbeys. Extensive roads were built without maps for guidance.

Field surveys, as we know them today, for the draughting of accurate maps depend on the development of precision theodolites and levels which were not available before the eighteenth century. Maps were required originally for military purposes, but it was after 1750 that the engineer surveys in Ireland, India and eastern North America were undertaken, although regional cartography has been attempted from the sixteenth century. The Society of Arts, from 1759 until 1801, offered a premium of one-hundred pounds for any original county survey upon a scale of one inch to a mile.

Colonel Sir Charles F. Close (1865–1952) says in *The Early Years of the Ordnance Survey* that 'the English county maps had no pretentions to accuracy; they were inferior, in this respect, to Rennell's recent maps of Bengal, and their errors cried aloud for correction'.

Between 1725 and 1740 General George Wade constructed 250 miles of military roads in Scotland and it was the difficulties he experienced followed by the defeat of Charles Edward, the Young Pretender, by the Duke of Cumberland at Culloden in 1746 that convinced the Deputy Quarter-Master-General, Lieutenant-General Watson, that a map of the Scottish Highlands was essential. This was the first suggestion that an Ordnance Survey should be carried out by the Army on the authority of the Government.

There followed wars with the American Colonies, who finally declared their independence in 1776, wars with France in 1778, with Spain in 1779 and with Holland in 1781, so until peace was signed in 1783 there was no time for large surveys. By this date, France had become more advanced in both geodetic and topographic surveys than the British Isles.

Jesse Ramsden had made his first dividing engine in 1766, and completed an improved automatic engine in 1775, capable of dividing and ruling circular metal scales to every 10 seconds of arc. By 1793, Edward Troughton had also completed a circular dividing engine capable of cutting 24 divisions a minute on a theodolite scale (sections 1.4 and 1.9).

At this time, the plane table had been in

use for at least 200 years, but it was the development of the optical theodolite, with engine-divided metal scales, which made it possible for rapid progress to be achieved in accurate surveying.

In 1784 an order was placed with Ramsden for a 36 in theodolite with a horizontal brass circle, divided into 10 minute spaces, which would be read by micrometer microscopes. This instrument, used between 1787 and 1853, is now on display at the Ordnance Survey headquarters in Southampton.

In 1765, General William Roy was appointed Surveyor-General of Coasts and Engineer for making and directing Military Surveys in Great Britain. He founded the National Survey, but the Duke of Richmond (1735–1806), Master-General of the Ordnance, made the survey possible and officially started the Trigonometical Survey in 1791.

The first Triangulation of Great Britain was built up over a period of 60 to 70 years during the nineteenth century. A network of large triangles, covering the British Isles, was selected from available observations and called the Principal Triangulation.

A new triangulation—the Retriangulation, to eliminate boundary discrepancies—was started in 1935 and developed since 1945 to provide a completely consistent National Grid network with modified Transverse Mercator coordinates originating at Lat. 49° North and Long. 2° West of Greenwich. For convenience, in applying this system to Great Britain, a false origin is employed and 400 km are added to all easting coordinates and 100 km subtracted from all northing coordinates. Rectangular coordinates quoted by the Ordnance Survey are thus related to a working origin Lat. 49° 46′ N, Long. 7° 33′ W of Greenwich, which is south-west of the Isles of Scilly. This ensures that coordinates of all points on the mainland of Great Britain are positive and less than 1000 km (Fig. 6.1.1).

The first-order stations of the primary retriangulation are about 30 to 50 km apart and this network is broken into second-order stations 8 to 12 km apart. Third-order stations are 4 to 7 km apart over the whole country except mountain and moorland areas. In

6.1.1

Diagram showing 100 km squares and the letters that designate them.

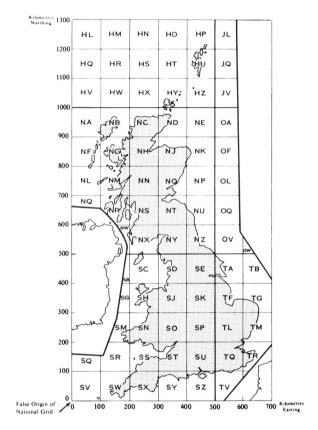

urban areas, where the scale of survey is 1 : 1250, the density of the control network is increased by fourth-order stations 1 to 2 km apart.

The Secretary of State for the Environment in Parliament on 19 February 1973 stated that:

The Ordnance Survey will continue to function as the central survey and mapping organisation in the public sector with the following aims:

(a) To produce and to maintain up-to-date basic surveys at 1 : 1250 for major urban areas and at 1 : 2500 or 1 : 10 000 for the remainder of the country.

(b) To make this survey information available at the basic scales in such forms as may be most appropriate to the needs of

168

users; and additionally in the case of the 1:10 000 scale to publish and maintain a uniform series of maps covering the whole country.

In order to maintain up-to-date basic surveys, the primary purpose of the national survey is to record change in the landscape as it occurs. The pace of map revision must match the rate of change, so conventional maps printed on paper no longer hold a monopoly position in the communication of topographic information. Large-scale surveys will become available on magnetic tape and maps will be miniaturised on microfilm.

The first priority involved concentration on the production of three series of large-scale maps. The 1:1250 (approximately 50 in to 1 mile) covering large urban areas, the 1:2500 (approximately 25 in to 1 mile) covering all cultivated areas and small towns, and the 1:10 000 for mountains and moorland. These are known as basic scales and all other map series are derived from the detail of the basic-scale maps. The 1:10 000 series maps are based on metric values. In terms of area, 1 cm^2 on the 1:10 000 map represents 1 ha (2·471 acres) on the ground.

Surveying consists of measuring the relative positions of natural or man-made features on the surface of the Earth and plotting them to form a map or plan. Geodetic surveying takes into account the curvature of the earth and involves spherical geometry. The Ordnance Survey of Great Britain is a geodetic survey. Plane surveying assumes that the area under consideration is a horizontal plane and can be used for surveys up to 250 km^2 in area, when the difference between a horizontal plane and the tangent to the Earth's surface is not significant. Plane surveying is used for the measurement of areas and plans for legal documents required by civil engineers, builders and town planners. Site surveys at scales of 1:200 and larger are generally undertaken by land surveying methods, but the combined theodolite and electronic distance-measuring instruments are replacing the separate theodolite and steel tape.

Levelling is concerned with the third dimension which is normal to the horizontal. The relative difference in height or elevation between various points on the Earth's surface, based on the Ordnance Datum, has been carried out over most of Great Britain. A system of cyclic levelling every 20 years, except moorland and mountain areas which are relevelled every 40 years, ensures a knowledge of subsidence or other changes in the crust of the Earth.

A benchmark is defined as a mark which has been determined by spirit levelling at a height which is relative to the Ordnance Datum. Benchmarks are derived from second- or third-order geodetic levelling and established at about 5 per square kilometre (km^{-2}) in rural areas and 30 to 40 km^{-2} in urban areas.

The Ordnance Datum is the mean level of the sea of Newlyn in Cornwall which was calculated from readings on an automatic tide gauge from 1915 to 1921.

6.2 Steel-tape surveying and plane-table equipment

Land-measuring chains were used by the Romans for the fixing of boundaries, and for setting out cities, aqueducts and roads. The link chain of wrought iron, introduced in the seventeenth century, has been replaced by steel-wire chains, 36% nickel–steel tapes up to 50 m in length of known coefficient of expansion, and optical squares for establishing right-angles.

A freely suspended invar–steel tape hangs in a catenary curve, but the required length is the distance along the chord. If the tensioning weight is the same as that at which the tape was calibrated, then no correction for sag is required as the zero marks give the standard length along the chord. If a tape has been standardised in the flat, then a negative correction must be applied, as the chord distance will always be less than the graduations on the tape.

An optical square can be used by a builder or surveyor for setting out angles of 90 or 180° to an accuracy of about 1 in 600. The square consists of two pentagonal glass prisms, each having a deflection angle of 90°, mounted in a casing which can be held in the hand.

In order to mark out a line perpendicular to a chain line through a point on a hedge or boundary, the observer holds the optical square and walks along the chain until the image of the boundary mark in the prism coincides with a ranging pole at the end of the chain (Fig. 6.2.1).

To find a point on a line joining two terminal poles, the observer walks across the line looking through the prism so that the pole on the left is seen in the upper prism and the pole on the right is seen in the lower prism. When these appear to be exactly in line, one over the other, the optical square is over the line joining the poles (Fig. 6.2.2).

The plane table is used to prepare a map or plan in the field and, except when contouring, without calculation. The technique has been

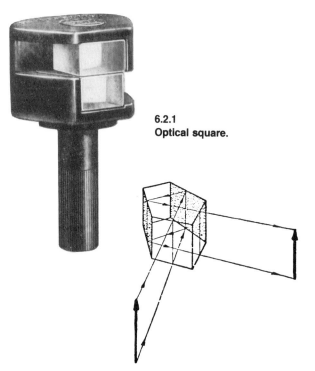

6.2.1
Optical square.

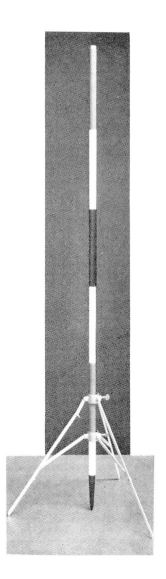

6.2.2
Ranging poles are either made of wood or light alloy tubing. They are painted red, white and black. A metal folding-pole support is used on hard ground or paving where poles cannot be driven into the ground.

used since the sixteenth century, but it was the Survey of India, in the eighteenth century, which revealed its possibilities. An alidade, or sighting rule, which may include a telescope and scale, must be accurately levelled on the table so that when sighting a distant object the angle of elevation or its tangent can be read and recorded (Fig. 6.2.3).

170

6.2.3
(*a*) Watts microptic alidade and plane-table equipment
including prismatic compass, bubble level and plummet
support. (*b*) General arrangement of microptic alidade
with (*left*) field of view of circle-reading eyepiece showing
sexagesimal angular scale (top) and vertical and
horizontal Beaman scales.

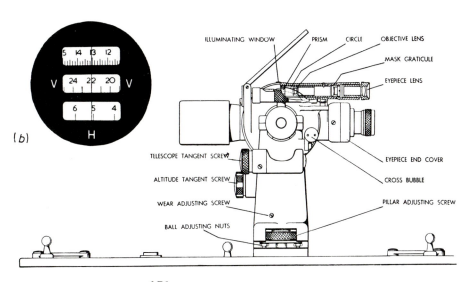

171

The Watts microptic alidade is a 15× telescope which can be levelled, by means of a two-minute cross-bubble, and has a screw-focused eyepiece with 1:100 stadia lines on a graticule. The altitude bubble has a hinged mirror reader so that all observations can be made without the observer changing his position. By means of the parallel rule, a plane can be produced directly so the scale must be known before work begins.

A plate-glass scale can be used for the measurement of distances between points on maps, plans or aerial photographs. Two magnifying scale readers are adjustable along guide rails for accurately setting on points and reading to the nearest 0·1 mm (Fig. 6.2.4).

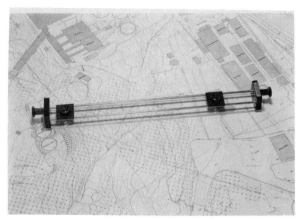

6.2.4
Measuring scale for surveying, cartography and photogrammetry.

6.3 Traverse surveying

The most important instrument for exact survey work is the theodolite, based on Kepler's design of telescope, which will measure very accurately both horizontal and vertical angles. The theodolite was invented by Leonard Digges in the sixteenth century and, following the first telescopic instrument completed by Jesse Ramsden in 1787 which did not have a vertical circle, has been continuously under development. A transit theodolite has a telescope and a vertical circle connected to it which can be revolved on its trunnion axis through 360°. Traditionally, the telescope image was inverted on theodolites and levels, because Kepler's design of telescope consisted essentially of two convex lenses. In order to use this type of telescope, the figures on staves must be inverted (Fig. 2.1.7). Modern instruments usually incorporate an erecting lens, without any loss of performance, and so this avoids possible reading errors (Fig. 2.1.8).

The Watts No. 1 microptic theodolite has a telescope magnification of 25× and a 3·8 cm aperture. The glass circles with chrome lines and figures are easily read direct to 20 seconds and by estimation to 5 seconds of arc. The optical plummet rotates with the instrument and is self-checking. Both circles are seen through the reading eyepiece, on the outside of the upright, which can be set to the most comfortable reading position independently of the telescope angle. The horizontal and vertical graduations appear in apertures marked H and V, respectively, and the micrometer scale, which is common to both circles, appears below the aperture. Each circle is observed in turn (Fig. 6.3.1).

This theodolite is detachable from the levelling base, so that it may be precisely interchanged with a target for sighting an optical plummet or index head up to a distance of 300 m for catenary taping (Fig. 6.3.2).

A traverse survey involves the measurement of angles between successive bearings of each line and the length of the lines. When a number of points, required by the survey, are joined these become the traverse lines. Wherever possible, the traverse lines are joined up to form a polygon, so that any errors which may have occurred can more easily be detected and corrected, and this is then called a closed traverse.

The interchangeability of instruments and accessories has revolutionised the art of traversing as, instead of plumbing the instrument or target over each traverse point, it is now commonplace for the three-tripod system to be used. Additional speed and convenience is achieved with an extra tripod which enables the theodolite and targets to be exchanged in order to progress from one station to the next

172

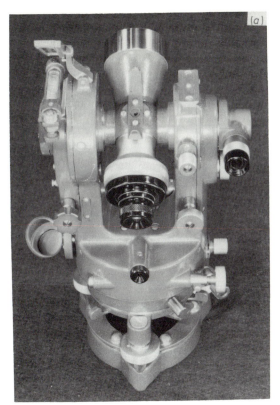

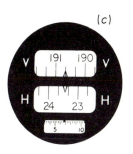

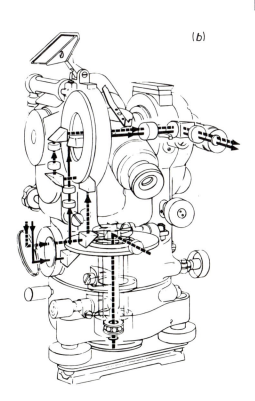

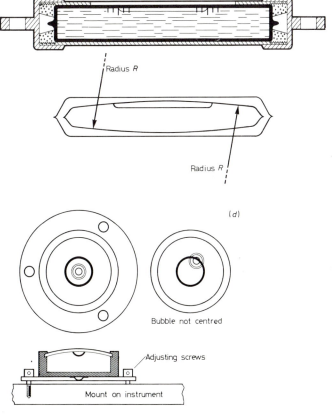

173

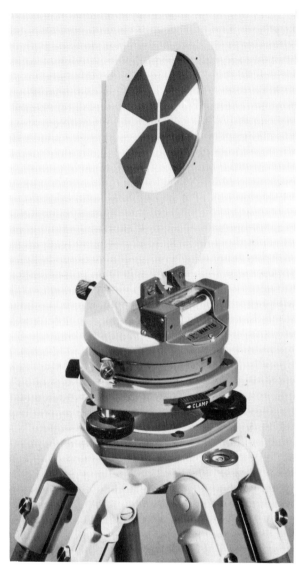

6.3.2
Target with optical plummet—a target which can be used at close range or up to 300 m. The optical plummet and bubble level are similar to those on the theodolite and the centre of the target is the same height, above the base, as the horizontal axis of the theodolite telescope.

without the need to mark each station on the ground. Theodolites, targets and plummets are made with exactly similar base fittings of strong construction which can be rapidly attached or removed from bases on the tripod heads.

The engineer's transit is a robust compact instrument which can be used for all forms of simple theodolite surveying, including tacheometry and solar observation, for preliminary reconnaissance and civil engineering surveys.

Linear and base-line measurements may be by catenary taping. Two people are needed to carry out the work with a surveyor to direct operations and record the details. A steel chain or invar measuring tape, together with a spring balance to tension up to calibrated value, a level and a thermometer are required. Temperature corrections are needed if ambient temperature is different from the tape calibration temperature.

Although angles are usually graduated in the sexagesimal system from 0 to 360°, with subdivisions in minutes and seconds of arc, sometimes gradations are specified in the centesimal system as 400 major divisions to a circle, and these units are known as grades. The grade is subdivided into five intervals each of 20 minutes and so, with this system, angles can be expressed as decimals of a circle.

The Cooke V22 scale-reading theodolite is suitable for civil engineering, mining surveying and tacheometric surveying. Traverse surveying, using the conventional three-tripod method, is simplified by a quickly detachable base (Fig. 6.3.3).

Horizontal and vertical circles are read through a single circle-reading eyepiece lying alongside the main telescope. The reading is direct to 30 seconds of arc and by estimation to 6 seconds through microscopes. The scale method of angle reading can be considerably faster than the micrometer method (Fig. 6.3.4).

The altitude spirit vial, essential for the precise determination of vertical angles, is entirely enclosed in the left-hand standard and insulated against temperature changes and transit shocks. The altitude vial is read by means of a prismatic coincidence-reading system which increases the setting accuracy by a factor of two. The V22 theodolite has an optical plummet, focusing from 1 m to infinity, for centring over a ground mark and is self-checking.

There are some theodolites in regular production which are direct-reading to 1 second

174

6.3.3
(*a*) Cooke V22 scale-reading theodolite mounted on levelling base and tripod. (*b*) Schematic cut-away diagram of Cooke V22 scale-reading theodolite. The horizontal and vertical circles are read through a single circle-reading eyepiece lying alongside the main telescope.

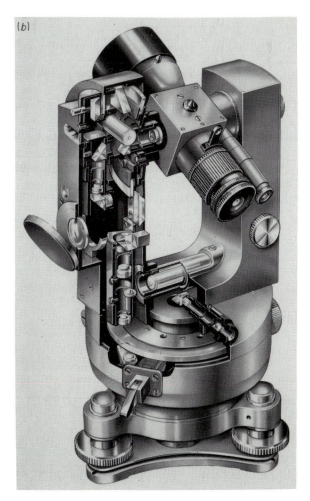

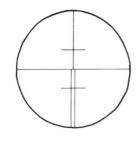

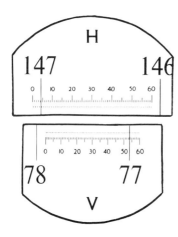

6.3.4
Telescope graticule with horizontal stadia lines (*upper diagram*) and scale reading from the eyepiece (*lower diagrams*). Scale reading: horizontal 147° 4′ 15″; vertical 77° 53′ 45″.

of arc or to a higher order of accuracy if required. In order to achieve this performance, it is essential. that observations are taken from diametrically opposed positions on the circular scale and the mean reading used to eliminate any centring errors.

The Wild T16 scale-reading general-purpose theodolite can be used for low-order triangulation, traversing, tacheometry, detail survey and field astronomy. Both horizontal and vertical circles are seen simultaneously and readings can be estimated to 0·1 minutes of arc. A red warning screen blocks out the vertical circle if the instrument is not level. This instrument can be used in conjunction with an infrared 'Distomat', a GAK1 north-seeking gyro, or a laser eyepiece (Fig. 6.3.5).

The Watts No. 2 microptic theodolite will focus down to 2 m but at 300 m will resolve intervals of 3 mm. The horizontal glass circle is 98 mm diameter and the vertical circle 76 mm diameter with direct reading to 1 second of arc (Fig. 6.3.6).

Usually, the levelling screw base is an integral part of the instrument but, if intended for use with three tripods, the upper part may be detached from the levelling base and exchanged for targets, optical plummets, or catenary taping index heads.

There are two adjustments for telescope focusing. The orthoscopic eyepiece must be adjusted so that the graticule lines are sharply in focus and, as this is a personal setting, it will vary from one individual to another. The second adjustment is to the internal negative lens so that the image to be observed is formed in the plane of the graticule. After removing the telescope cap, the telescope should be directed to the sky and the eyepiece adjusted until the graticule lines appear sharp and black. Then direct the telescope to the target and focus, by means of the knurled sleeve on the telescope tube, until the mark appears stationary relative to the diaphragm. When the head is moved slightly from one side to the other there should be no parallax. The term parallax is used if there is movement of the image of one stationary object with respect to the image of another stationary object, when the eye is moved sideways. If the images of the target and graticule lines move relative to one another then there is parallax and an adjustment is necessary.

The line of sight of the telescope must be at right-angles to the trunnion axis and any departure from this condition is known as collimation error. In order to obtain optimum accuracy, when measuring horizontal angles, the circle should be set to read zero by adjusting the milled head on the levelling screw base. Starting at 'Station A', a round of clockwise readings should be taken closing on 'Station A'. Reverse the telescope and take a round of readings in an anticlockwise direction. Now take a second, third and fourth

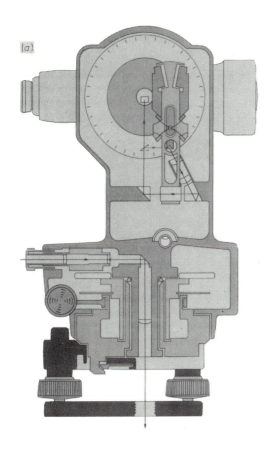

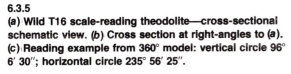

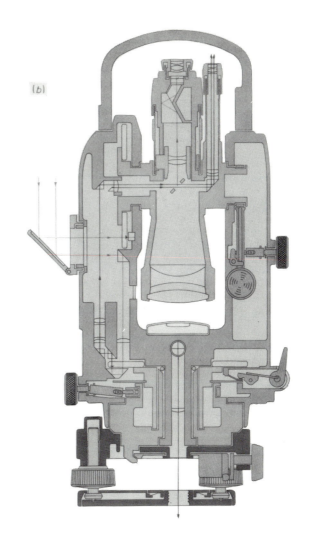

6.3.5
(a) Wild T16 scale-reading theodolite—cross-sectional schematic view. (b) Cross section at right-angles to (a). (c) Reading example from 360° model: vertical circle 96° 6′ 30″; horizontal circle 235° 56′ 25″.

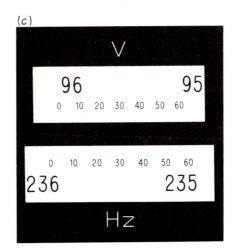

series of observations, the second commencing at 90° with a micrometer reading of 2·5 minutes, the third at 180° with a micrometer reading of 5 minutes and the fourth at 270° with a micrometer reading of 7·5 minutes. This will ensure elimination of errors in collimation, and transit axis adjustment, and also that the effects of any gradation or micrometer run error are considerably reduced.

The telescope graticule is engraved above and below the horizontal line with stadia

6.3.6
Watts No. 2 microptic theodolite.

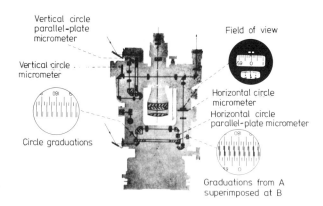

6.3.7
A theodolite with glass circles. The diagram shows the optical system for the simultaneous reading of both the vertical and the horizontal circles, although each circle has an independent light path. It also shows the principle of bisection readings and the automatic compensation of eccentricity errors by reading both sides of a circle at once.

The altitude bubble, completely enclosed within the upright, is observed by the coincidence method which eliminates errors due to meniscus, parallax or changes in the length of the air bell. In the 'split bubble' system, the observer sees the two ends of the bubble side by side and has to bring them into coincidence so as to form a smooth curve. The observing prism is rotatable so that observation can be made from any head position (Fig. 6.3.8).

6.3.8
By means of an optical system, both ends of the bubble are seen as split images side by side in the eyepiece. When the telescope is tilted, the two images move in opposite directions.

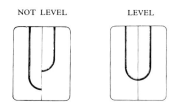

marks which subtend an angle of 1:100. If the line of sight is horizontal, the distance from the centre of the instrument to the levelling staff is given by multiplying the staff intercept by 100.

Each circle has its own micrometer which consists of a parallel-sided glass block situated in the light path from one side to the other side of the circle. Each block can be tilted, by rotating a cam, and the tilting of the block causes an optical shift of the double lines. The spindle on which the cam is mounted also carries a small circle graduated with 600 divisions (Fig. 6.3.7).

178

For very precise setting out of industrial installations, it is possible to use an auto-collimating telescope in which the conventional internal focusing system is combined with an autocollimating eyepiece with a filament lamp (L) and semi-reflector (S) (Fig. 6.3.9).

The Wild T0 compass theodolite can be used either for observing and setting out magnetic bearings, or as a normal theodolite for measuring angles. It is particularly useful for rapid traversing, based on magnetic bearings, in areas where visibility is limited and traverse legs are likely to be short. The compass circle is lowered onto its pivot by a spring lever. Horizontal and vertical circle readings are made to the nearest minute of arc (Fig. 6.3.10).

The Wild GAK1 gyro attachment when fitted to the top of a theodolite, such as the Wild T16 scale-reading theodolite, orientates to True North within ±30 seconds of arc in

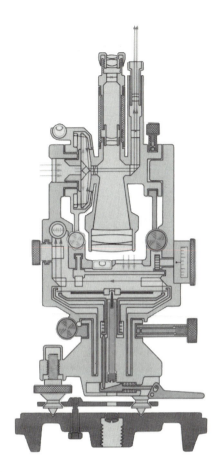

6.3.10
Wild T0 compass theodolite—schematic cross section.

6.3.9
Schematic illustration of autocollimation and autoreflection. (a) Watts theodolite telescope with autocollimation and autoreflection. (b) Graticule patterns as seen during autocollimation. (c) Graticule patterns in coincidence. (d) Autoreflection field of view.

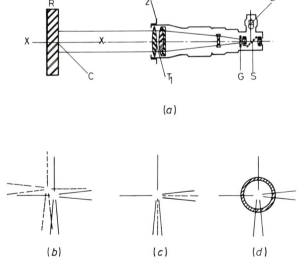

about 20 min of time providing the initial setting is within ±30° of True North. When used in mining, a gyro attachment can establish the absolute azimuth at the foot of a shaft and only one plumbing is required for positioning. The gyro may be used to determine the directions for laying out long narrow constructions, such as canals, roads and railways, or the orientation of radar transmitters and airfield wind vanes.

When a theodolite has been levelled up, the gyro motor, which is suspended on a thin metal tape, hangs like a plumb-bob so that its spin axis is always horizontal. The gyro rotor,

179

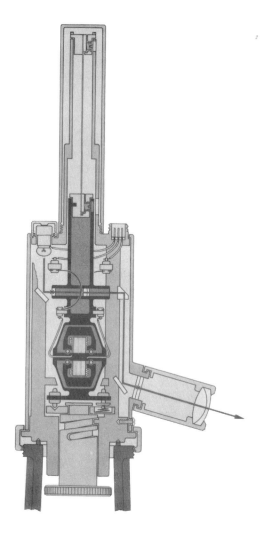

when rotating at 22 000 rpm, tries to maintain its initial random spinning in space but is pulled out of its original plane by the rotation of the Earth. The gyro reacts to this interference by precession until it takes up a position in the True North meridian. The gyro oscillations about this plane can be observed, through the observation tube, and so the theodolite can be set at the mean of the oscillating movements of the gyro axis (Fig. 6.3.11).

6.3.11
Wild GAK1 gyro attachment—schematic cross section.

6.4 Tacheometry

Optical measurement of heights and distances can be determined directly from instrument readings, and so if necessary the steel-taping operation may be eliminated. This is of importance if the land is cut by ravines, river valleys, or standing crops where direct measurement would be inaccurate or difficult to achieve. Tacheometry is a rapid method compared with measurement by tapes and the results can be of similar accuracy. Tacheometric survey detail is plotted manually, using a polar coordinatograph, and becomes control detail. The remaining detail is completed by a surveyor, using visual alignments and intersections, with taped short-distance measurements.

Tacheometry, or stadia surveying, involves the determination of distance by measuring the angular subtense of a known traverse length at a distant point. Telescopes of all surveying instruments have graticules which, in addition to cross-lines, have stadia markings. Usually, these are marked above and below the centre line so that the angle subtended between them, when viewed against an object, is 1:100 from what is known as the anallactic point (which may be the instrument centre). Sometimes, stadia markings to the left and right of the centre line may be provided. Graticules are used either as scales for direct measurement, or as setting lines where measurements are made on a scale included in the instrument (Fig. 6.4.1).

If the line of sight is horizontal, the distance from the centre of the instrument to the levelling staff or rod is given by multiplying the staff intercept by 100. The equivalent horizontal and vertical distances on inclined readings are given by the following expressions:

$$\text{horizontal distance} = 100S \cos^2 V$$
$$\text{vertical distance} = 100S \tfrac{1}{2} \sin 2V$$

where S is the stadia reading, and V is the angle of inclination.

The stadia principle was first described in 1778 by William Green, who used two fixed wires, but James Watt used a tacheometer of his own construction in 1771.

Instead of using two graticule lines a precise distance apart, it is possible to direct a theodolite at a staff or rod, make two pointings, and measure the small subtended angle. If a levelling staff held vertically is used, this is known as the tangential system. If a horizontal bar of fixed length is used, this is called the subtense system (Fig. 6.4.2).

6.4.1
Graticule patterns for theodolite and level telescopes. The short lines are very accurately positioned as they are stadia markings for distance measurement.

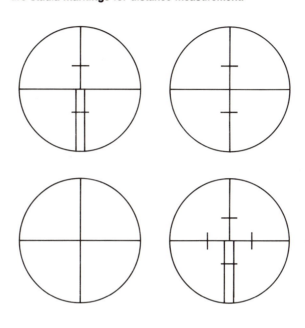

6.4.2
The Watts 2 m subtense bar has targets supported on invar–steel rods which have a low coefficient of expansion, so that distance measurements are unaffected by wide variations in temperature.

When using the subtense bar, it is possible to determine distance by measuring the horizontal angle subtended by the targets located at each end of the calibrated bar. A conversion table will show distances against subtended angles. If a theodolite reading direct to 1 second of arc is used, then a precision of 1:10 000 can be easily achieved (Fig. 6.4.3).

One of the advantages of the subtense bar method of measurement is the ability to vary accuracy according to the requirements of particular surveys. Instead of using a fixed stadia angle, giving a constant factor of 100, and measuring the intercept cut by this angle, the subtense bar has a constant intercept and a variable parallactic angle to be measured. A modern subtense bar will have easily observed aiming marks, a levelling up bubble, a sighting device so that the staff-man can direct the bar towards the observer, and a collimating device enabling the observer to check for himself that the bar is set up correctly.

Another method of distance measurement makes use of a tacheometer prism attached to the front of a theodolite telescope, together with a special horizontal staff and Vernier scale. It is possible to achieve an accuracy of 1 in 3000, and readings of range correct to 1 cm, with this equipment (Fig. 6.4.4).

The point of the staff rod is placed on the station mark and the points of the wooden legs are trodden into the ground. The prism attachment, known as a Richards' optical wedge and invented by R. Bosshardt in 1923, is placed on the objective end of the telescope. The micrometer control is rotated until the direct and deviated images of the staff are superimposed on each other. If the telescope is pointed and the cross-hairs are sighted on the zero of the staff, and the wedge is placed in position, then the cross-hairs will move to a different position depending on the deflection factor of the wedge. If the wedge is constructed to refract a total deflection of 34' 22·6", then this will give a tangent of exactly 1:100 so the amount of movement of the cross-hairs, multiplied by 100, will give the distance from the telescope to the staff. If the

6.4.3
The subtense bar is mounted on a tripod at one end of the distance to be measured. A theodolite is set up at the other end. The angle subtended by the two targets when sighted in the telescope of the theodolite is measured on the horizontal circle. The distance between the 2 m subtense bar and the theodolite is then obtained from the relationship.

$$D = \tfrac{1}{2}b \cot\tfrac{1}{2}\theta.$$

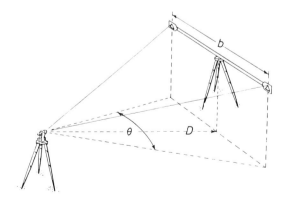

Accuracies of measurements by the direct method (assuming an error of 1 second of arc in reading the subtense angle) arc given in the table:

Error† in direct measurement of distance for 2 m bar with 1 second error in subtense angle.

Distance (m)	Error (m)	Proportionate error
15	0·00054	1 in 28 000
30	0·0022	1 in 14 000
45	0·005	1 in 9000
60	0·009	1 in 6500
90	0·019	1 in 4700
120	0·035	1 in 3400
150	0·054	1 in 2800
300	0·22	1 in 1300

†These errors are derived from the formula:

$$E_D = \frac{D^2}{b} \times E_\theta$$

where E_D is the error in the measured distance and E_θ is the error in the measured angle.

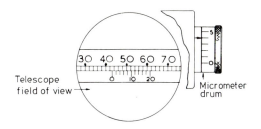

6.4.5
View of tacheometric stave as seen through the telescope with prism mounted on the objective. An optical micrometer is incorporated in the prism attachment on the telescope. Reading: staff, 42 m; Vernier, 1·2 m; micrometer, 0·04 m; so total distance is 43·24 m.

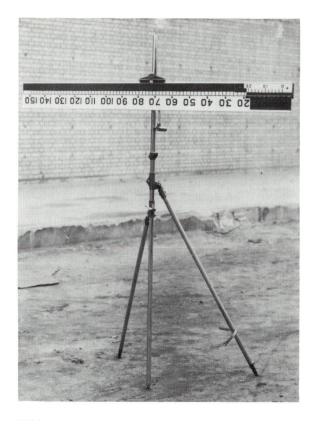

6.4.4
Watts tacheometric stave which is engine-divided from 20 to 150 cm. The first 10 cm are occupied by the Vernier which enables readings to be made to one-tenth of the main scale division.

wedge covers only half the lens, then only half the image will be moved. If the main scale is viewed directly, and the Vernier is viewed through the wedge, then the amount of movement will be read with Vernier accuracy. Further accuracy can be achieved with a parallel-plate micrometer. The main scale divisions give the range in multiples of 2 m and the lower of the two main scale divisions, between which the Vernier zero lies, is recorded. To this is added the Vernier reading, which gives an addition in multiples of 20 cm, and exact coincidence between a main scale graduation and the Vernier is achieved by rotating the parallel-plate micrometer knob on the telescope (Fig. 6.4.5).

Distances from 300 to 7000 m can be measured with a Wild TM2 coincidence range-finder. The theoretical accuracy of this instrument varies from ±2 m at 500 m to ±194 m at 6000 m range. The optics give a large erect image of the target with another small rectangular inverted partial image in the centre. The coincidence setting screw is used to displace the inverted image until it coincides exactly with the features on the erect image. The distance to the target is then read through the left eyepiece (Fig. 6.4.6).

Electromagnetic distance-measuring (E.D.M.) equipment has been developed for long-distance measurement, using light waves or microwaves based on the time taken for the waves to travel to a reflector and back again.

Although various types of electronic processing equipment are used, all E.D.M. devices are based on phase comparison to determine distance. A high-frequency carrier wave is modulated by a low-frequency measuring pattern wave, transmitted to and then reflected at the remote station back to the instrument where the phases of outgoing and returning waves are compared.

A difference in phase between the two waves represents the fraction of a wavelength by which the return distance exceeds an integral number of complete wavelengths. The

(a)

6.4.6
(a) Wild TM2 coincidence rangefinder for use on land.
(b) View through the eyepiece of a coincidence
rangefinder illustrating the inverted partial image
in the centre.

(b)

number of complete wavelengths is unknown until measurements are repeated using different frequency waves and then a unique pattern of phase differences will be found which are applicable to within a certain range of distances. This technique is similar to the fraction-coincidence method of calculating the thickness of gauge blocks by a gauge interferometer as described in section 8.8.

The Kern Mekometer ME3000 is an electro-optical distance-measuring instrument with high resolution at short and medium distances. In contrast to conventional amplitude modulation, the polarisation of an optical carrier is modulated. In this particular instrument, polarisation modulation is used as a means of measurement. The high modulating frequency of about 500 MHz provides a high absolute distance resolution and permits an optical phase measurement by

means of a variable light path instead of an electrical phase meter (Fig. 6.4.7).

The modulation wavelength is fixed by the resonance of a relatively small microwave resonator containing air under atmospheric conditions. Thereby, the distance scale is affected by atmospheric changes much less than with instruments operating with fixed modulation frequencies and with variable modulation wavelengths (Fig. 6.4.8).

Under favourable atmospheric conditions, distances up to 1·5 km can be measured employing one reflector, while with three reflectors distances up 2·5 km can be obtained. A complete measurement with five frequencies for distances over 1 km requires 2–3 min. Relative measurement (e.g. deformation measurements) require only measurements with the first frequency, when the time needed is less than 2 min.

6.4.7
Kern Mekometer ME3000 electro-optical distance-measuring instrument: S, station; Z, target; E, reference plane within the distance meter for phase comparison between transmitted and received waves; R, reference plane for the reflection of the wave transmitted by the distance meter; a, addition constant; e, distance-meter component of addition constant; r, reflector component of addition constant; λ, modulation wavelength; Ø⌐ fraction to be measured of a whole wavelength of modulation. Distance is calculated from

$$L = \tfrac{1}{2}n\lambda + \tfrac{1}{2}\varnothing$$
$$a = e + r$$
$$D = L + a.$$

The addition constant a applies to measuring equipment consisting of distance meter and reflector. The components e and r are only auxiliary quantities. The components r of all ME3000 reflectors are equal within a few hundredths of a millimetre, so these reflectors can be interchanged.

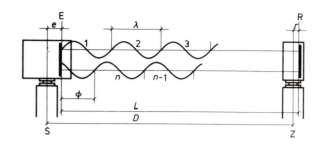

6.4.8
An instrument operating with fixed modulation frequencies and with variable modulation wavelengths. *Optical system:* 1, xenon flash tube; 2, 3, crossed polarisation filters; 2, polarisation filter; 3, analyser; 4, modulator crystal; 5, demodulating crystal; 6, reflector; 7, variable light path; 8, photomultiplier. *Modulator unit:* 4, modulation crystal; 5, demodulation crystal; 9, modulator; 10, reference resonator; 11, local oscillator for adjustment of the five modulation wavelengths necessary for a complete distance measurement; 12, resonance detector and frequency control unit; 13, switch for adjustment of the five modulation waves and for computer control; 14, frequency tuner; 15, frequency tuning indicator; 16, desiccant; 17, vent. *Phase meter unit:* 3, analyser; 7, variable light path; 8, photomultiplier; 18, phase detector and amplifier control; 19, handwheel for the light path (7); 20, signal strength display; 21, phase adjustment indicator; 22, computer with distance display. *Power supply unit:* 23, power source; 24, transformer, distributor and timing pulse generator.

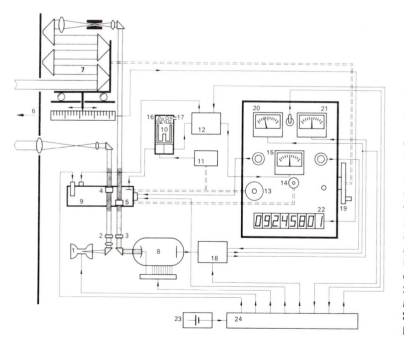

185

7 Aerial Surveying and Photogrammetry

7.1 Introduction

Aerial photography, other than air survey, includes low-altitude oblique photographs of factories, civil works, power plants or building sites which may be required for commercial or publicity purposes. Low oblique photography is of value in determining the limits of flooded areas and in preparing schemes for land drainage (Fig. 4.6.4).

Air survey is concerned with mapping of the Earth's surface by means of accurate vertical or high oblique photography. Aerial photographs enable accurate topographical information to be obtained of undeveloped areas of the world, such as swamps, impenetrable forests and large tracts of country that may offer prospects for development. In well developed countries, air survey can be used to advantage for recording the progress of roads, railways, goods yards or docks under construction. Photographs may serve purposes other than ordinary mapping, such as forestry or agricultural surveys, geological and archaeological investigations.

Aerial photographic surveying is well adapted for the revision of existing maps, at the three basic scales which have been established by ground survey methods. New surveys should have a number of 'ground control' points in the area fixed by ground topographical surveying.

The aerial survey is divided into four main stages which are:

(*a*) Obtaining the aerial photographs.
(*b*) Coordinating the plan and height points necessary for controlling the plotting process.
(*c*) Plotting the map detail.
(*d*) Field completion by ground survey.

Vertical photographs are taken with the camera axis vertical when the exposure is made. In practice, the aircraft tilts during flight within 1 and 3°, and this results in a tilted photograph which must be allowed for in the processing to ensure precise photogrammetry.

Oblique aerial photographs including the apparent horizon, which may be a broken line if the terrain is mountainous, are known as high oblique photographs. Low oblique photographs do not include the horizon and are usually for pictorial purposes or for presenting a detailed view of non-technical committees. They are easier to interpret than vertical photographs because they present the usual view from the top of a hill or building.

High oblique photographs have the advantage that the view of the stereoscopic model is natural, as seen by an observer, and so enables well trained operators to plot quickly

186

and accurately. Often, high oblique and vertical photographs are taken at one exposure to save time and expense. A disadvantage of oblique photographs is the distortion, so, usually, for large- and medium-scale surveys, vertical photographs are used, but high oblique photographs are suitable for small-scale maps (Fig. 7.1.1).

A high cloud layer and clear atmosphere is essential for good aerial photography, as a small cloud and its shadow will usually cover part of the photograph which is vital for plotting.

Photogrammetry is concerned with the measurement of photographs and it was D. F. J. Arago (1786–1853), of the French Academy of Science, in 1851, who demonstrated the application of photography to topography. At about the same time, Porro (1853) established his principle of observation through lenses and Meydenbauer (1858) commenced research on the application of terrestrial photogrammetry to architecture in the design of monuments.

In 1852, Aimé Laussedat, an officer of the Corps of Engineers of the French Army, thought of the idea of utilising terrestrial photography for compiling topographic maps. By replacing the plane table with phototheodolites, the image points on terrestrial photographs could be intersected in space. In 1894,

Colonel von Hubl, chief of the topographic section of the Austrian Geographic Institute, adopted the methods of Laussedat for high mountain surveys.

The stereocomparator, originally developed in 1903, was transformed into a drawing instrument in 1908 by Lieutenant von Orel who was attached to the Geographic Institute of Vienna. His stereoautograph traced a map point-by-point automatically, as with modern plotting machines. The invention by H. E. Sainte Claire Deville (1818–81) of the first automatic plotting instrument, for photogrammetry in Canada, opened the field for photographs obtained from phototheodolites.

The most accurate results from plotting are obtained from vertical photographs, which have been taken with the axis of the camera pointing vertically down, in a 60% overlap in the direction of flight between successive photographs.

Adjacent strips with common detail should have an overlap of about 30% to provide complete stereoscopic cover in spite of deviations from course and aircraft lateral tilt. The length of flight for each strip is usually about 10–15 km, as longer strips require ground control points to ensure accurate navigation. Ground control points are essential for contouring and must be established before or after the photography has been completed (Fig. 7.1.2).

The advantage of selecting ground control points from a photograph is to ensure positive identification and favourable location of the

7.1.1
Relative areas covered by vertical and oblique photographs from the same altitude.

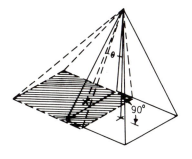

7.1.2
Overlap of vertical photographs.

points. If the points must be marked on the ground before aerial photography, then they should be easy to see by the photographer, and preferably be in the form of a white cross large enough to be identified from the aircraft at the time of exposing the photographs (Fig. 7.1.3).

Because the angle of reflection of light received from the Earth by an observer in an aeroplane is very different from the angle of reflection at ground level, the appearance of objects will usually be different. For example, a field of corn will appear light from the air on a windy day, but if the air is calm it will be dark due to the millions of shadows cast by individual stalks and leaves. Ploughed black earth after rain may appear white from the air, because it will reflect sunlight, but this depends on the direction of the light.

Photographs should always be viewed with light from the same direction as the sunlight, during exposure of the film, or depressions may be misinterpreted as elevations. Shadows of high objects will indicate the correct orientation for the viewing light.

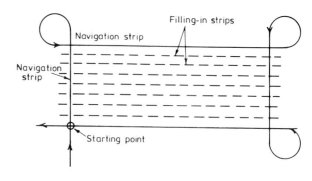

If the photographs do not contain hilly country, and are not tilted, then contouring of the map involves the use of a topographical stereoscope fitted with movable parallactic grids. Stereoscopy plays an important part in air survey plotting and the interpretation of aerial photographs.

7.2 Stereoscopic vision

Stereoscopy is the science for viewing objects in three dimensions which enables us visually to determine relative heights and distances. Stereoscopic vision is essential to the photogrammetrist who must understand the mechanics and mathematics of observations. If any object within the field of both eyes is viewed, first with one eye and then with the other, it will be seen to move back and forth sideways against the background.

Vision with both eyes is binocular vision, and the fusion of two perspective images before the eyes into a single spatial model in the brain is physiological, and this phenomena is known as stereoscopic vision. The crystalline lens of a human eye is a converging lens of variable focus and the alteration of focusing distance, or accommodation, is obtained by changing the curvature of the lens (Fig. 2.1.1).

Two points A and B can be viewed and images seen at a, a' and b, b'. As a result of the variable focus of the eye, the nodal points N move according to the distance of the point of focus. In normal binocular vision, the lines of sight from the two eyes intersect on the object where the vision is concentrated. The angles between the line of sight for the two eyes, intersecting at A and B, are called parallactic angles for the two points and, for distinct vision at about 250 mm, the limiting upper value is about 16°. The lower limiting angle ranges from 10 to 20 seconds of arc, at a distance of about 700 m, depending on the eye separation, which is usually about 65 mm (Fig. 7.2.1).

In order to be able to see depth beyond 700 m, the eye separation must be increased by means of binoculars or a rangefinder. Alternatively, two photographs must be taken at separate camera stations (on the ground, from the ends of a fixed base, or from an air

The lines of sight from two eyes intersect on the object where vision is concentrated. This is known as binocular vision.

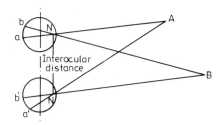

The fused images of b and b' appear to be closer to the observer than the fused images of a and a' (see text).

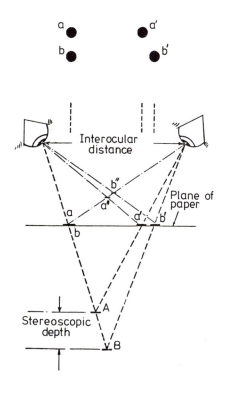

base) and the photographs viewed with a stereoscope (Fig. 2.1.13).

If four dots are drawn on a piece of paper, with a spacing less than the interocular distance, and a person looks at a point halfway between the dots a and a', they appear to fuse together and two dots will be seen. When the two pairs of dots are fused together, the dot formed by b and b' appears to float in space below the dot formed from a and a' (Fig. 7.2.2).

If the procedure is reversed, so that one is looking at objects A and B through a transparent screen, then images will appear on the screen at a, a', and b, b'. From the stereoscopic images of two points in space, their relative heights may be obtained.

M. Hotine has reported that objects between infinity and 3 m can be fused by normal accommodation of the eyes, and H. G. Fourcade has shown that objects 300 and 332 mm from the eyes can be clearly focused at the same time. In each case, the convergence is about 70 minutes of arc, so if the difference of convergence of a pair of dots is much greater then it will not be possible to fuse them if they are in the same line or close together. Within a field of 2° there must not be a local change of more than 70 minutes of arc in order to be able to fuse points together (Fig. 7.2.3).

At the heights appropriate for aerial photography, the stereoscopic relief obtained by a pair of human eyes is negligible so overlapping photographs from a series of exposures are placed in a stereoscope. The effect is

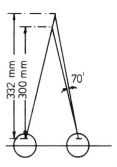

Fourcade has shown that when objects at a distance of 300 mm from the eyes are clearly focused, then objects 332 mm away can also be seen clearly. In each case, the difference in convergence is about 70 minutes of arc. It appears that, within a field of about 2°, no greater range than this can be dealt with if all the stereoscopic picture is to be seen in fusion.

the same as if the scene was viewed by a giant, who had an eye base equal to the distance between the two positions of the aircraft when the exposures were made. The length of the air base is usually about 15 000 times the length of the eye base, which for most people averages 65 mm, representing 975 m between each photograph for aerial photography at a height of 5000 m. The Helmholtz stereoscope, originally without magnification, was first produced in 1857 (Fig. 7.2.4).

Stereoscopic parallax has an important bearing on the determination of relative heights and in setting aerial photographs to their correct orientation. When a train is moving at high speed, a person sees telegraph poles near the track pass rapidly with respect to the window frames. A row of fence posts 300 m away moves less rapidly, and buildings 5 km away appear to move much more slowly than the fence posts or telegraph poles. The relative movement to the window frame is known as parallax, and the closer objects have a greater parallax than the distant objects.

When applied to aerial photography, the images of the high points on the terrain will move more rapidly than the lowest parts in the valley, so the higher levels have a greater parallax than the lower elevations. If S and S' are two successive camera stations at which vertical photographs have been taken, then, if the air base B and flying height H are known, it is possible closely to estimate the heights of the ground line above the datum plane (Fig. 7.2.5).

In forestry studies, aerial photographs can be used for classifying the nature of timber and determining tree heights. The variations in growth of trees, depending on soil and rock conditions, can be easily recognised from aerial photographs (Fig. 7.2.6).

The stereoscope, in which the pair of photographs are observed and fused together, is fitted with a pair of glass plates which have a number of marks or, alternatively, a single pair of marks which can be set anywhere in the overlapping photograph area. A scale is provided for measuring coordinates both along the flight line and those perpendicular to the air base.

The pair of dots in Fig. 7.2.2 represent a crude form of floating mark which illustrates how a change of spacing alters the stereoscopic depth. If two corresponding points are selected on adjacent photographs, then, if a pair of marks are used as an indicator and they have the same spacing, they should appear at the same level. If they are less

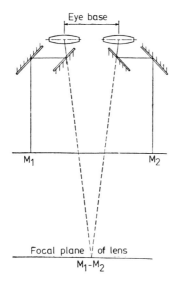

7.2.4
The Helmholtz stereoscope is a reflecting stereoscope, usually made of mirrors or of mirrors in combination with lenses.

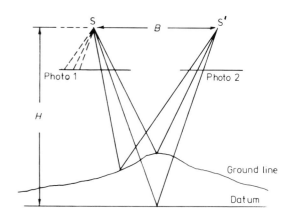

7.2.5
Parallax displacements due to relief.

190

Parallax applied to the measurement of tree heights.

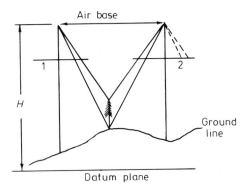

For topographical mapping, it is essential to make precise photogrammetric measurements, so calibrated cameras with wide-angle lenses of short focal length are needed to obtain a large base/height ratio. The larger the base/height ratio, the greater will be the parallactic angle and accuracy of stereoscopic measurement (Fig. 7.2.8).

Flying heights vary from a few hundred metres, with large-scale helicopter photography, to several hundred kilometres, if satellites are used to carry the camera. Usually, for topographical mapping, the flying height will vary between 500 and 3000 m, but sometimes higher altitudes are used depending on the terrain to be surveyed and the map scale.

Because air density, and also the refractive index of air, decreases with increased altitude, the light rays reflected from the ground are bent by atmospheric refraction. So, an error is introduced on the aerial photographs and the image coordinates must be corrected during processing (Fig. 7.2.9).

Radial distortion will be caused by the Earth's curvature, and it may be necessary to compensate for image distortions on vertical photographs taken from high-flying aircraft. These distortions become more severe as radial distances from the point on the photograph beneath the exposure station to the object point increases (Fig. 7.2.10).

For accurate surveys, it is essential to take account of the tilt and height of the camera at the time of exposure. A compound stereoscope can be used to determine the angular

widely spaced they will appear above, and if more widely spaced they will be below the photograph. Marks are usually in the form of dots, triangles, crosses or arrows. Because the image of each point on the terrain has moved between successive photographic exposures, then each image has a slightly different parallax to the next image and this makes stereoscopic viewing possible (Fig. 7.2.7).

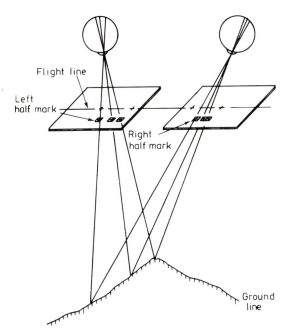

The principle of the floating mark.

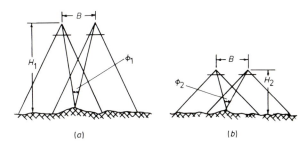

Parallactic angles increase with increasing base/height ratios, i.e. since $B/H_2 > B/H_1$ we can see that $\phi_2 > \phi_1$.

7.2.9
Atmospheric refraction in aerial photography.

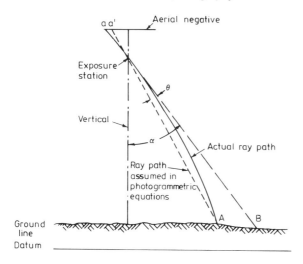

If a tilted photograph is rectified, then it is equivalent to a vertical photograph and can be enlarged or reduced by any desired magnification to bring a series of photographs to the same scale at a particular elevation.

Because aerial photographs must be taken in rapid succession, large rolls of film are needed in the camera. The negative (which lies in the image space behind the lens) produces a reversal of all object points on the terrain. For the purpose of examination in a plotting machine, a printed positive on glass is required, known as a diapositive, and this is similar to a photograph.

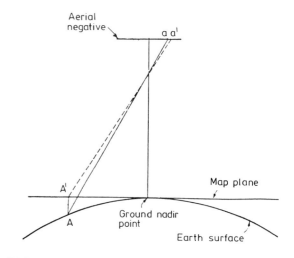

7.2.10
Distortions in aerial photography due to Earth curvature.

measurements from which the relative positions and tilts of the camera can be computed, providing three ground control points appear in the series of photographs. Calculations can be avoided by the use of automatic plotting machines.

Radial line plotting is based on the photographs being vertical, so if tilt exists then this principle does not apply. Usually, tilt does not exceed 1° and, as the direction of tilt is random, some effects tend to cancel out and photo control will usually limit major errors.

7.3 Topographical surveys and cartography

Hunting Surveys Ltd have kindly provided the photographs and basic information for this section. The range of the company's activities include aerial photography, photogrammetric mapping, traverse and level networks, land and hydrographic studies, engineering and marine geophysics, and these have been developed over more than 50 years experience.

In many parts of the world, water has a value greater than any other natural resource and surveys are required for:

(a) The conservation of water by construction of dams and reservoirs.

(b) The conveyance of water by aqueduct, pipelines and canal systems.

(c) The spreading of water over irrigable land.

All these projects need large-scale topographical plans as well as longitudinal, cross-sectional and digital volumetric data for capacity or earthwork calculations.

The requirements for pipeline engineering may be concerned with networks to collect or distribute water, sewage, oil or gas, and require an area survey. Another type of survey is required for pipeline routes transporting oil, gas, water or products over long distances.

Where there are no medium-scale maps or charts, it is necessary to carry out aerial photography and mapping, and also field surveys and preliminary hydrographic surveys, in order to select the general alignment of a pipeline route.

For engineering purposes, conventional light aircraft fitted with an automatic pilot, are suitable for carrying a remote sensor which is usually a photographic camera using panchromatic film, but may be:

(a) Magnetometers or gamma-ray spectrometers for oil and mineral exploration.

(b) Infrared thermal sensors, for 8 to 12 μm wavelength, which detect temperature variations of $\pm 0\cdot5\,°C$ at the ground surface. These are useful for groundwater studies, exploration for sources of geothermal power, the location of leaks in pipes, and sources of pollution.

(c) Microwave systems which transmit a polarised signal and record the reflections from the terrain onto an oscillating line scanner. Microwave radar can penetrate haze and cloud and, because it is independent of sunlight, is serviceable at night. This technique is of value in the humid tropics where conventional photography is difficult because of the atmospheric conditions.

Radar imagery can be made into mosaics at 1:100 000 scale and can be used in a similar way to the conventional photomosaic. Hunting Surveys Ltd have covered the whole of Nicaragua and large areas of Brazil, Colombia and Indonesia by radar mosaics which are being used for geological exploration and natural resources surveys.

Control surveys are required for oil and mineral exploration which must define concession boundaries. Electronic instruments enable distances to be measured as quickly and as accurately as angles (Fig. 7.3.1).

All electromagnetic distance measurement depends on physical line-of-sight conditions between stations but, whereas light-wave instruments require clear visibility, microwave equipment will operate successfully through haze, cloud, rain and thin trees. Microwave E.D.M. instruments are mainly used for long-range measurements from 3 to 40 km with an accuracy of ± 2 cm $\pm$ 5 parts per million. Measurements require a surveyor and instrument at each end of the line. A speech facility is provided which is invaluable for organising the survey programme over these long distances. Some instruments combine a theodolite with electromagnetic distance measurement in a single piece of equipment (Fig. 7.3.2).

Spirit levelling for determining the height of benchmarks is time-consuming and costly over long distances. Accuracies of 1 cm km^{-1} may be considered normal but 4 mm km^{-1} can be obtained with precision levels. Aneroid barometers measure pressure differences very accurately but, when converting these into differences of height, one assumes stable atmospheric conditions. If simultaneous readings are taken with several barometers at different stations, the accuracy of height measured is not likely to be better than ± 2 m, but this may be good enough for undeveloped country. Hunting Surveys Ltd used helicopters in difficult areas of Ghana and Nigeria to establish photocontrol and barometric levelling.

All photogrammetric mapping relies on a certain amount of ground survey work to establish the scale and height datum of the stereogram or model, and the points selected for this purpose must be identifiable on the ground and on the aerial photograph. The scale of the photography is determined by the flying height and focal length of the camera lens, but economy usually demands that the smallest acceptable photoscale is adopted. Weather conditions may overrule this judgment, as it may be possible to fly at 650 m under light cloud, but be impossible to obtain good photographs at a higher altitude.

All modern survey cameras use film with a negative size of 23 cm × 23 cm and usually have lenses of 150 mm focal length. Longer focal length lenses may be used for mapping urban detail, and for making mosaics. Shorter focal lengths give greater height accuracy and enable small-scale photography to be flown at lower altitudes, although there will be more distortion and less resolution at the edges.

Height accuracy from the map plotting machine is determined by the flying height of the photography. Providing the ground surface is clearly visible, and there are identified height control points in the corners of the photographs, then spot heights can be measured to an accuracy of 2 or 3 parts in 10 000 of the flying height.

If an aerial survey for urban mapping is photographed at 375 m (1250 ft), with a 150 mm (6 in) lens, the scale will be 1:2500

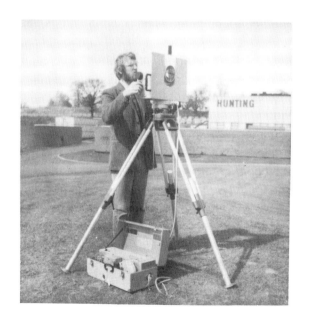

7.3.3
Infrastructure. Aerial photography, surveying and mapping are essential techniques in the design and development of all communications and services. For preliminary route selection, photomosaics and small-scale maps are needed. Final design demands much larger-scale photography and maps. Permanent ground markers are established by land surveying methods to provide control for setting out construction. Hunting Surveys Ltd provides all these services for the development of highways, railways, canals, water supplies, drainage, telecommunications, pipelines, power transmission and airports.

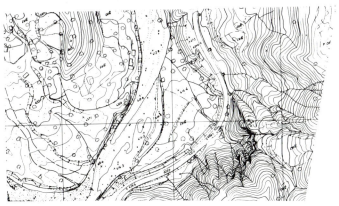

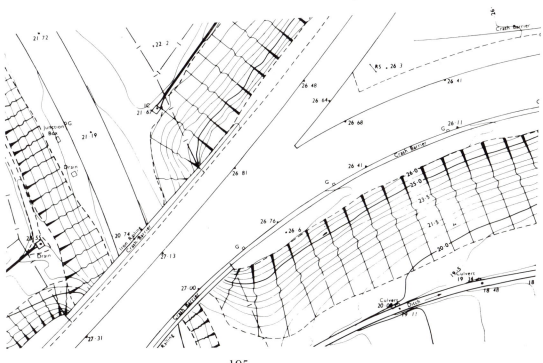

and this will be suitable for 1:500 scale mapping after enlargement. Spot heights of 10 cm (0·33 ft) and contours at 30 cm (1 ft), on ground not covered by vegetation, are achieved and widely used for engineering public utilities such as drainage, water and gas supply.

Pipeline feasibility studies in unmapped areas overseas may be flown at 6000 m (20 000 ft) giving a scale of 1:40 000 which, with a sparse scattering of ground control points to give strip mapping at 1:10 000 scale, would enable contours at 10 m vertical intervals to be established.

Engineers should make more use of aerial photographs and photomosaics which can be produced at low cost. A set of overlapping vertical photographs joined together on a backing board give an aerial view, but no uniformity of scale, and this is known as an uncontrolled mosaic. If rectified photographs are brought to approximately the same scale and fitted with ground control points, then the mosaic is said to be controlled. Some of the distortion inherent in aerial photography can be removed during printing, but in hilly terrain the scale errors and distortion are considerable. Cartography is the oldest and most popular method of presenting the survey data since maps, plans and profiles are easily understood (Fig. 7.3.3).

Aerial surveys for potential reservoir sites have been widely used based on 1:10 000 scale of photography and existing Ordnance Survey maps. One of the requirements may be the provision of depth and capacity tables which are obtained by digitising contours, or reading off cross sections, on the plotting instrument. By the same method, the amount of material extracted from a quarry can be calculated.

A technique recently developed is known as orthophotography. This method produces photographic plans with the same accuracy as a conventional line map, without scale differences or distortions inherent in a photographic mosaic. This form of mapping is useful in deserts and open scrubland as it will give a better overall picture than a line map.

Orthophotography has been used for land-reclamation studies and, increasingly, at the planning stages for road construction and the siting of new factories. A contoured orthophotograph map was made for a reservoir survey in Indonesia where conventional plotting of innumerable terraced rice fields would have been slow and costly.

The majority of plotting instruments are of the analogue type where the stereo pair of survey photographs, in the form of positive prints on glass, are placed in carriers which give the same mechanical reconstruction as the original aerial camera system. The diapositives are viewed through a binocular which gives a 6× to 10× enlargement and, when the plate carriers are correctly orientated, the terrain will be seen stereoscopically in three dimensions (Fig. 7.3.4).

Within the monocular system of each eyepiece, a small reference mark in engraved and, when viewed stereoscopically, the reference marks fuse together when placed on a feature of the stereoscopic model. By means of handwheels, and a footwheel to adjust for height, it is possible to traverse over the terrain, tracing detail which can be transferred to the plotting table through a train of gears at the required scale.

In order to establish the exact spatial condition, i.e. that applied at the moment of exposure of each photograph, a first-order plotting machine enables an operator to use the reference marks to correct distortions in the photographs. Adjustments of position and inclination of the photographs within the instrument are achieved by the introduction of lateral or fore-and-aft tilt, swing, relative height and horizontal separation.

The same reference marks are used to follow and record the features required on the survey map. By systematically following all of the topographic details on the stereo model, whilst the plotting pencil is in contact with the paper, the map is drawn and numerical values can be read or automatically recorded. In many cases, the data are required only in numerical form and no graphical plot is prepared.

7.3.4
Orthophotography and analogue type of plotting instrument for viewing terrain stereoscopically in three dimensions.

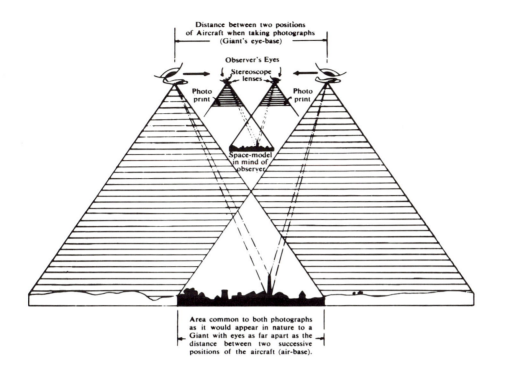

Distance between two positions of Aircraft when taking photographs (Giant's eye-base)

Observer's Eyes

Stereoscope lenses

Photo print

Photo print

Space-model in mind of observer

Area common to both photographs as it would appear in nature to a Giant with eyes as far apart as the distance between two successive positions of the aircraft (air-base).

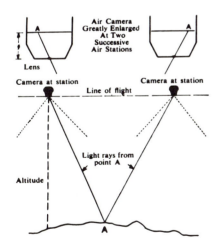

Air Camera Greatly Enlarged At Two Successive Air Stations

Lens

A

A

Camera at station

Camera at station

Line of flight

Light rays from point A

Altitude

A

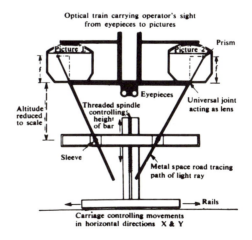

Optical train carrying operator's sight from eyepieces to pictures

Picture 1

Picture 2

Prism

f

Eyepieces

Altitude reduced to scale

Threaded spindle controlling height of bar

Universal joint acting as lens

Sleeve

Metal space road tracing path of light ray

Rails

Carriage controlling movements in horizontal directions X & Y

197

7.4 Aerial photography

Most large-scale air survey photographs are obtained through a wide-angle lens with a focal length of 150 mm having an angular field of about 90°, such as the Wild Universal Aviogon I or II.

These lenses were designed and patented by Wild in the early 1960s and gave a much better marginal illumination than any previous lens. From pure geometrical principles, the light from a normal lens falls off as the fourth power of the cosine of the semi-angle. By introducing marginal aberrations in the front half, it was possible to increase marginal light to the third power of the cosine of the semi-angle, and correct the aberrations in the rear half. The lens design is complicated and is based on the principle of using large negative meniscus lenses with the convex side on the outside (Figs 7.4.1 and 7.4.2).

Photographs of urban areas and forests and orthophotographs require a longer focal length lens such as the 300 mm Wild Aviotar II (Fig. 7.4.3).

The lenses are supplied with a 420 nm haze filter and a 525 nm dark-yellow filter, each filter having a graded-density layer for

7.4.2
The Wild Universal Aviogon II ($f/4$, $f = 15$ cm) is particularly well suited for use under poor light conditions and for large-scale low-altitude photographic flights.

7.4.1
The Wild Universal Aviogon I ($f/5.6$, $f = 15$ cm) is used for medium-scale and large-scale topographic mapping, cadastral survey, aerial triangulation and also for the production of orthophotographs.

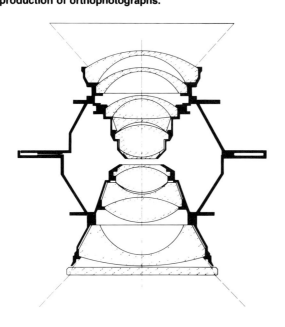

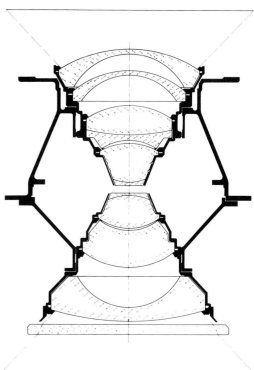

198

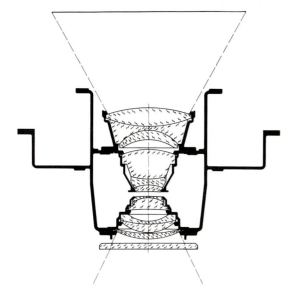

7.4.3
The Wild Aviotar II (*f*/4, *f* = 30 cm) is a high-speed lens suitable for large-scale mapping of urban areas and forestry, for orthophotographs and photomosaics.

7.4.4
The Universal Aviogon I and II, and also the Aviotar II, are inserted directly in the camera mount and can be interchanged in flight. The image format of all lens cones is 23 cm × 23 cm.

compensating the light fall-off of the lens, which makes it possible to produce visible or infrared photographs. A circular opening of 400 mm diameter in the camera mount receives the lens cone which includes a variable internal iris diaphragm. A longitudinal and lateral tilt correction of up to ±5° can be introduced during flight in order to level the camera manually or by remote control via electric sensors (Fig. 7.4.4).

The Wild Aviophot camera can be controlled from a central position and direct attention by an operator is needed only for changing film cassettes, lens cones or filters. A floor opening of 400 mm diameter is needed for the camera, and also a 120 mm diameter hole for the viewfinder, if the floor thickness is 150 mm. The viewfinder is used for navigation and for orienting the camera during flight. The 110° viewing angle enables almost the entire area covered by a super-wide-angle photograph to be viewed.

A clock with sweep-second hand, an altimeter with dial reading and a note panel are projected onto a data strip outside the picture frame for imaging on the film within the 230 mm × 230 mm format. The spools used in the standard cassettes will accommodate up to 150 m of film, and a mechanical gauge indicates the length on the spool.

An overlap regulator releases a series of photographs automatically with exposure intervals depending on the flying height, ground speed, field angle of the lens and desired forward overlap of the photographs. Depending on their purpose, a wide range of overlap ratios are needed from 20 to 90% of the frame size.

The exposure meter consists of a sensor under the aircraft and an exposure calculator placed next to the control unit, with a spectral sensitivity through the visible spectrum into the near-infrared region. The exposure time, as a function of the aperture selected and effective film speed, is automatically transmitted to the camera (Fig. 7.4.5).

Carl Zeiss Oberkochen also supply aerial mapping cameras having an automatic exposure cycle, film-flattening, film-transport and interval cycling facilities (Fig. 7.4.6).

7.4.5
Wild RC10 Aviophot cameras include interchangeable
lens cones, individual cassettes and automatic exposure
control with preselected aperture.

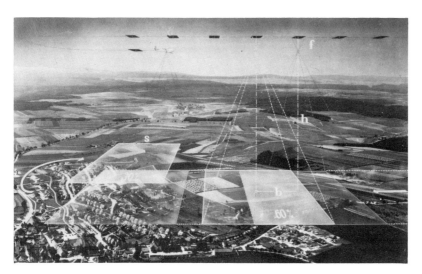

7.4.6
Systematic aerial photography of
an area with automatic control of
exposure cycle, film flattening,
film transport and interval
cycling.

Zeiss lenses are available for the following principal uses (see also Fig. 7.4.7).

Lens	Angular field (deg)	Description
85 mm S-Pleogon	125	Large-area small-scale mapping for flights below cloud cover
153 mm Pleogon	93	General aerotriangulation and large-scale mapping
210 mm Toparon	75	Intermediate angle where 305 mm focal length is unsuitable
305 mm Topar	56	Standard lens for aerial mosaics, orthophotographic maps and city maps
610 mm Telikon	30	Narrow angle for high-altitude photography and orthophotographic maps of urban areas with high-rise buildings

The camera has a rotating-disc shutter with four continuously rotating discs, with a maximum speed of 1/1000 s. The film magazine has a capacity of up to 150 m of polyester-based film, which can be suitable for panchromatic, infrared or colour photography (Fig. 7.4.8).

The Zeiss system of interchangeable camera bodies allows any of seven cameras to be used with the same suspension mount, film magazine, navigation sensor and navigation telescope (Fig. 7.4.9). The navigation sensor is an optical measuring instrument for determining angular speed and drift which is required in order to control the aerial mapping camera (Fig. 7.4.10).

Aerial survey cameras must be calibrated in

7.4.7
Carl Zeiss aerial survey lenses.

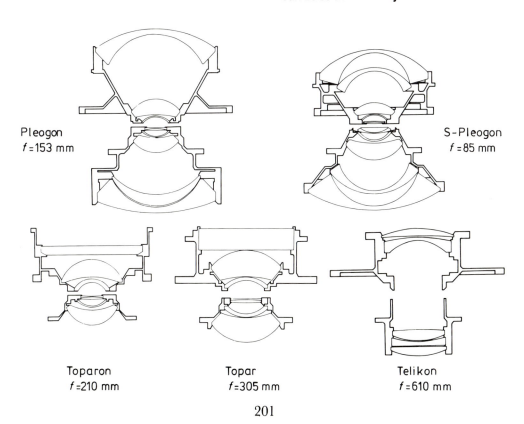

Pleogon
f = 153 mm

S-Pleogon
f = 85 mm

Toparon
f = 210 mm

Topar
f = 305 mm

Telikon
f = 610 mm

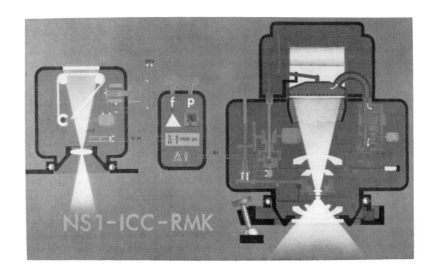

7.4.8
**Schematic diagram of a Carl
Zeiss aerial survey camera.**

a test fixture to ensure that the plane of the film during exposure is perpendicular to the optical axis. The point at which the principal axis meets the plane of the film must be fixed with an autocollimator and the principal distance must be measured when the lens is fitted to the camera, allowing for shrinkage of the film base. Camera adjustments must be possible during calibration to enable the principal point to be accurately located. High precision in the camera and lenses is essential for good-quality cartography.

7.4.9
Zeiss camera with navigation sensor and navigation telescope.

7.4.10
Navigation sensor with ground-glass screen for binocular viewing.

7.5 Stereoplotting instruments

First-order plotting instruments, suitable for large-scale surveys of maximum precision, are capable of making three-dimensional models of terrain from aerial photographs. Accurate plotting and heights of features can be achieved directly without calculations or the determination of coordinate points. Two or more successive photographs must be positioned relatively to each other to produce a geometrical model similar to the terrain, and then oriented absolutely with reference to ground points to establish the coordinates.

Second-order instruments are not capable of such large-scale plotting but provide exact solutions to the problems of relative and absolute orientation.

An alternative digital approach to stereoplotting, using orthophotographs, is based on an analysis of the coordinates which are derived mathematically by a digital computer.

Wild A10 Autograph stereoplotter

The analogue Autograph instrument, with electronic coordinate registration by digitisers, is designed to use the simple and proven principle of mechanical projection by means of space rods of constant length. The plotting table of area 1100 mm × 1400 mm will permit the plotting from edge to edge of models, formed from 230 mm × 230 mm photographs, without moving the manuscript (Fig. 7.5.1).

The bridge frame carries the two-projection-centre universal joints, separated by a fixed distance, and the projectors. In each projector, three principal-distance columns carry a frame for the picture carrier turntable and a carriage for the observation telescope. Each space rod moves a tie rod, in the plane parallel to the picture frame, and at its outer extremity is fixed the observation telescope which contains the measuring mark (Fig. 7.5.2).

The two projectors accept negatives or diapositives on film or glass up to a 230 mm × 230 mm image format. The photographs are centred on picture carriers, engraved with

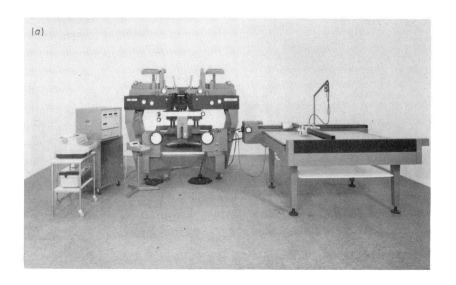

7.5.1
(a) Wild A10 Autograph plotting
instrument for the direct graphic
plotting of aerial photographs.
(b) Wild Aviotab TA plotting table
with digital servo drives for use
with Aviomap plotting
instruments and direct graphic
plotting.

fiducial marks, outside the Autograph on an accessory lightbox. For correction of radial distortion in the camera lens, aspheric correction plates engraved with fiducial marks can be used instead of the standard picture carriers.

The correction for Earth curvature and atmospheric refraction is achieved mechanically, from a large spherical segment on the base frame, by a lever system and reduction ratio controls.

The wide-angle viewing optics have a high resolving power over the entire field of view and the orthogonal viewing conveys a correct

204

7.5.2
Schematic diagram of the A10 Autograph.

spatial impression throughout the stereoscopic model.

Circular jet-black measuring marks, of 0·04 mm in the image plane, are an integral part of the viewing telescopes and always move parallel to the planes of the pictures. When plotting photographs, the floating mark is guided in plan by the two handwheels and in elevation with the foot disc which is coupled to a height counter (Fig. 7.5.3).

Interchangeable gears provide direct height readings from a mechanical counter, and plotting at all conventional map scales on the matt glass working surface of the table.

Carl Zeiss Oberkochen E3 stereoplotter

The Planicart stereoplotter has an integral tracing table and a three-dimensional compound-slide system in the model area for the unobstructed preparation of map manuscripts with a high vertical accuracy (Fig. 7.5.4).

With the aid of an external tracing table, a secondary magnification from model to map is made possible by the use of change gears which eliminate any restriction on the choice

7.5.3
Schematic diagram of the optical system with light shade and headrest opened.

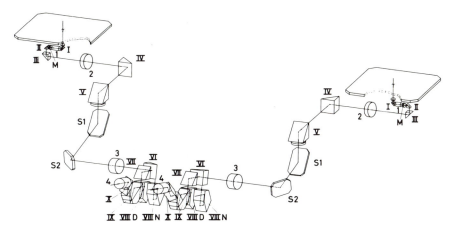

205

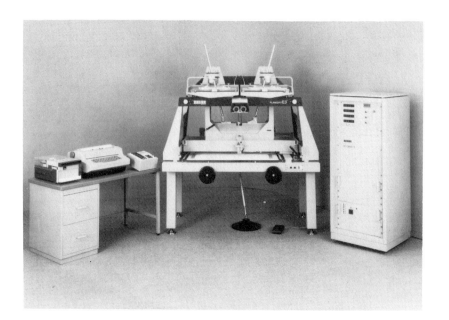

of economical photograph scales caused by specified flight heights (Fig. 7.5.5).

The optical system makes provision for eyesight correction, image rotation and a measuring mark for stereoscopic focusing (Fig. 7.5.6).

The Planicart stereoplotter is used for compiling charts and line maps at medium and large scales. When used for plotting points for aerotriangulation and digital mapping, an electronic recording unit, for the automatic triggering of preset increments of travel or time, can be connected. On-line computer-supported plotting with a desk calculator is also possible.

The Zeiss Planicart is not suitable for profiling orthoprojection work, but a simpler type of machine has been achieved by limiting the facilities to general mapping work.

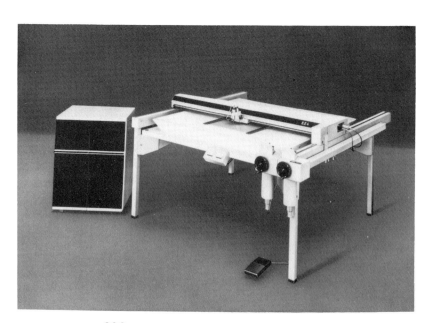

7.5.5.
Zeiss external tracing table with luminous surface and 120 cm × 120 cm plotting range.

206

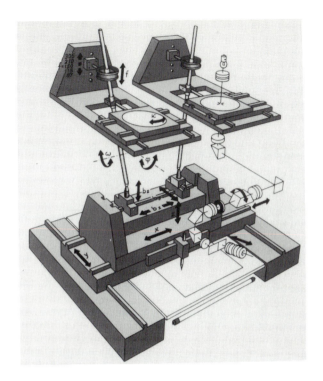

7.5.6
Zeiss Planicart: principle of projection.

Kern PG2 stereoplotter

Stereo-restitution instruments which are used for the production and revision of topographic maps, as well as large-scale plans, must be versatile and have a high accuracy. Manuscript and operational elements should be within easy reach of the operator. Photographic transparencies and paper prints up to size 230 mm × 230 mm from all types of camera with focal lengths between 85 and 172 mm can be accommodated. The PG2 plotting table is rigidly connected to the instrument. Orientation and plotting are based on customary photogrammetric methods, so that no special personnel training is necessary (Fig. 7.5.7).

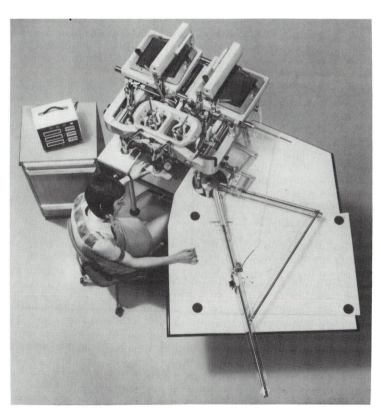

7.5.7
Kern PG2 with pantograph which enables various sizes of map enlargements to be drawn.

207

The photography is observed orthogonally so that secure guidance of the measuring mark is assured. The stereo image can be varied by the operator to 2×, 4× or 8× magnification. Elevations can be read directly at any model scale on a dial gauge which is attached to the base carriage (Fig. 7.5.8).

Corrections for Earth curvature can be made to elevations without affecting the parallax, and adjusted for any flying height. In the model space, conventional rods have been used for the reconstruction of the projected rays of light. The effect of camera tilt in the photography is corrected by adjustments to the space-rod systems instead of tilting the restitution projector (Fig. 7.5.9).

Because there are no rotating elements in the observation system, a relatively simple optical design is possible. The exit pupils of the oculars are 35 mm from the eyepiece, so that operators with spectacles can observe without inconvenience.

7.5.8
Kern PG2 main frame with adjustable supports and base carriage.

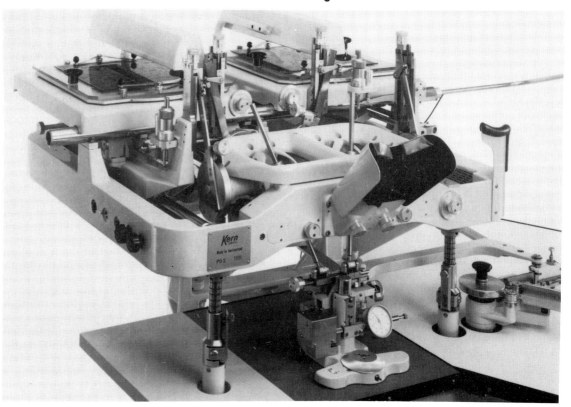

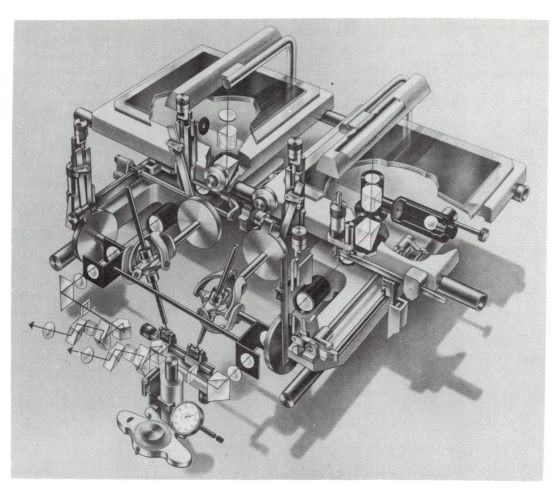

7.5.9
Design principles of Kern PG2.

Kern PG3 stereoplotter

The Kern PG3 large-scale plotter is designed for super-wide-angle photography and aerial triangulation based on the measurement of models. Camera lens focal length from 85 to 310 mm can be used in this machine (Fig. 7.5.10).

The PG3 basic design is very similar to the PG2 which uses the analogue system of mechanical reproduction from bundles of projected light rays. Scanning is carried out by means of two handwheels and a foot disc, as is usual for large-scale precision plotters. As the tilt elements are corrected in the space rods, a very simple observing system is used which does not require any rotating mirrors or prisms.

209

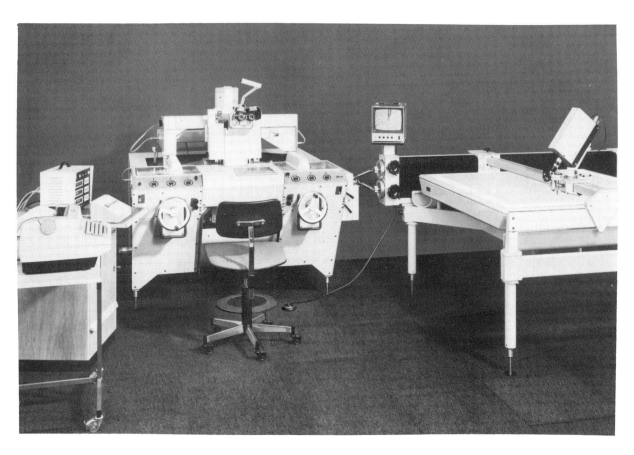

7.5.10
Kern PG3 plotter for large-scale maps.

The shape of the luminous measuring mark can be varied, as it consists of a ring, with outer diameter of 120 µm, and a concentric dot of 50 µm separated by a small empty zone. The illumination light can be switched off for either dot or ring so there is a choice of three different marks representing a dot, a ring or a combination of the two, which appears as a large dot.

A closed-circuit television system is available for close-up vision around the pencil on the drawing table. Glass scales can be used on the television screen for measuring cross sections, as a profiloscope. For aerial triangulation, cadastral surveys, or volume determination, counters and shaft encoders on the axes can be read visually or recorded on tape.

210

7.6 Orthophotographic cartography

Orthophotography is photographic cartography of the photogrammetric model in which a map, with all points to the same scale, is projected onto a photosensitive emulsion.

Aerial photography of uneven terrain can be rapidly transformed into an orthophotographic map by continuous exposure through a scanning slit, including corrections for the projection distance. The height of the ground is obtained from a stereoplotter.

Clarity of photographic detail can be combined with the horizontal accuracy of a graphical line map without any limitations because of ground relief.

Wild Avioplan OR1

Differential rectifications along the scan, and scale adjustment perpendicular to the scan direction are carried out simultaneously. This procedure ensures a perfect join between neighbouring scans independently of the ground slope perpendicular to the scan direction (Fig. 7.6.1).

Digital object data are supplied to the process controller of the Avioplan OR1 in the form of image coordinates, which may be measured directly in some stereoplotters or stored on magnetic tape for off-line operation.

For off-line operation, the image coordinates need not be measured directly in a stereoplotter but can be derived from arbitrarily distributed object coordinates by means of a computer program.

Photographs of long focal length can be used as input material, having a 20% overlap, with digital object data derived from photographs of shorter focal length and optimum base/height ratio (Fig. 7.2.8).

Matra Orthosfom 9300

Automatic orthophotographic plotting consists of making a map from a digitised terrain profile. Off-line digitised profiles can be used for automatic counter-line tracing, generation of profiles from digitised contour lines, or for updating of maps (Fig. 7.6.2).

For true on-line orthophotography, two complementary instruments are needed, consisting of a photoplotter and an orthophotographic unit. A variable enlargement ratio of 2·5 to 4 is possible between the exposure scale and the orthophotograph scale (Fig. 7.6.3).

7.6.1

Schematic optical diagram of the Wild Avioplan OR1 for orthophotographs.

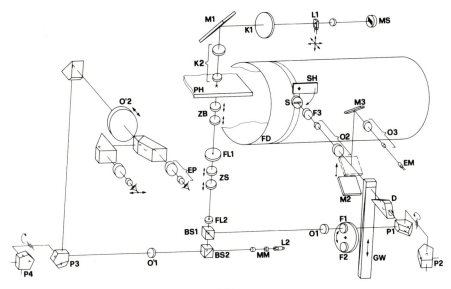

7.6.2
Automatic instrument for the production of orthophotographic maps from digitised profiles of the terrain.

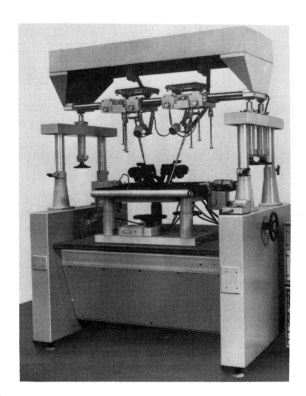

7.6.3
Matra Orthosfom 9300 for true 'on-line' orthophotography.

7.7 Stereoscopic measurement ———————————

Second-order instruments must provide exact solutions to particular problems of photogrammetry.

Matra analytical stereoplotter

The Traster 77 is an analytical stereoplotter with an accuracy at least equal to that of first-order instruments, but with superior performance characteristics and flexibility. Mathematical corrections can be made to various parameters such as distortion of lenses, atmospheric refraction, curvature of the Earth or film deformation. Geodetic projection in plan, graphical and digital photogrammetry can be performed.

Each of the photographs is projected onto the screen in polarised light. Stereoscopy is obtained by observing the screen with a pair of glasses with Polaroid lenses crossed at 90° and with an orientation corresponding exactly to the orientation of identical filters inserted in the optical projection path of each of the photographic images.

The screen display system used for the stereoscopic model permits total head freedom, and the table arrangement facilitates drawing without changing visual accommodation or moving from the work station. A conversational device gives the operator instructions on the cathode ray tube.

The high precision of this instrument permits aerotriangulation as on the best stereocomparator (Fig. 7.7.1).

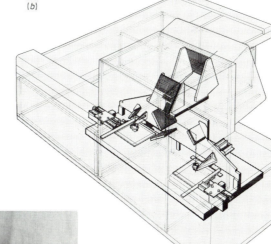

7.7.1
(a) Matra Traster 77 analytical stereoplotter. (b) Matra Traster 77 analytical stereoplotter for aerotriangulation. Viewing of the high-resolution polarised light on the screen is through spectacle polarisers which make it possible for several operators to see and discuss the images.

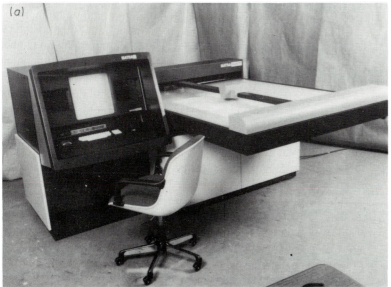

213

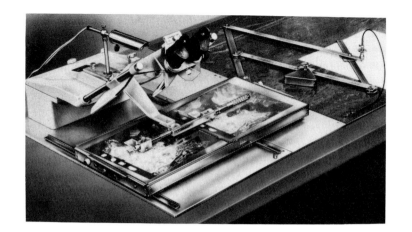

7.7.2
Zeiss Stereopret interpretation and drawing instrument.

Zeiss Stereopret

Simple stereoscopic measuring and tracing instruments are very popular for the interpretation of aerial photographs. Wherever the primary requirement is for point measurement, for which the high accuracy of stereoplotters is not needed, it is logical to connect a simple measuring instrument to a desk calculator or minicomputer. The image coordinates measured may then be used to compute ground coordinates such as distances, level differences, direction angles, slope, area or volume (Fig. 7.7.2).

Zeiss G2 Stereocord

An electronic unit can be used to digitise the coordinates measured on aerial photographs, terrestrial stereograms or non-stereoscopic photographs. Near-vertical or horizontal photography, with roughly parallel axes, can be plotted with the basic software (Fig. 7.7.3).

7.7.3
Zeiss G2 Stereocord stereoscopic measuring and tracing instrument.

214

8 Engineering Metrology

8.1 Introduction

Large-scale engineering projects could not have been undertaken by the ancient civilisations of Mesopotamia and Egypt without an understanding of the fundamentals of measurement. Sights, mirrors for setting squareness and alignment by in-line pinholes have been used for many centuries. The right-angle between a plumb-line and a water level was used to check the verticality of walls, whilst other angles could be made by subtraction or addition.

The royal Egyptian cubit standard of length was established by Amenophis I (1559–1539 B.C.), but in England, 2600 years later, Henry I (1068–1135) standardised the yard, which was revised in the reign of Elizabeth I (1588) as an end standard, and this was in use until 1824. The International Bureau of Weights and Measures (B.I.P.M.) was established in 1875, and a standard metre, a line standard made of 90% platinum and 10% iridium, was constructed in 1889.

The demand for accurate dimensions has increased rapidly from a thousandth of an inch (25 µm) during the nineteenth century to one-tenth of a thousandth of an inch (2·5 µm) or less from modern precision machinery. The means of measurement necessarily must be of at least ten times higher accuracy, in order to establish control over production, so optical instruments have been developed capable of this high performance.

As micrometres and seconds of arc are now normal measurement units for optical instrument manufacture, it is useful to consider their small size. A micrometre (1 µm) is to one metre (1 m) as 1 m is to 1000 km, and 1 second of arc is the gradient of 1 m in approximately 200 km. Because of the small dimensions required to be measured by optical instruments, care must be taken to control ambient conditions of temperature, air pressure and humidity. Details on this subject are given in *Dividing, ruling and mask-making* (section 10.4).

Alignment telescopes, autocollimators and interferometers have, since 1960, had the assistance of the laser with its intense coherent monochromatic light source. Laser instruments are used for engineering construction, alignment of tunnels, pipe laying, dredge guidance, offshore positioning and distance measuring.

The development and production of minicomputers and microprocessors at a low cost has made it possible to incorporate automatic calculators, with digital readouts, in the design of optical measuring instruments. The optical head as a primary source of information has, in some equipment, become part of an overall system incorporating electronic technology.

215

8.2 Observation telescopes

Small internal-focusing telescopes are used for observing scales on machine tools, the position of slides and workpieces on machinery, the progress of machining operations and for precise alignment applications. The steel tube of the telescope is usually chromium-plated and ground for direct mounting in V-shapes or tooling fixtures, when an accuracy of 75 μm in 15 m can be achieved. With an optical micrometer, small linear errors of alignment to 25 μm can be measured (Fig. 8.2.1). The basic design of a Hilger and Watts telescope is similar to a Keplerian telescope (Fig. 2.1.8) with an achromatic objective lens, a movable positive lens for focusing the rays from objects at various distances on the plane of a graticule, and an adjustable eyepiece.

Telescope eyepieces must be capable of adjustment to suit the particular eyesight of different observers. The dioptre is a unit of measure for eyesight, and is defined as the reciprocal of the focal length of a lens in metres, so a lens of 50 cm focal length will have a power of 2 dioptres. In order that an observer can focus when using a telescope, the eyepiece must compensate for defects in his vision over a range from about +6 to −6 dioptres. The power is often engraved as a scale on the eyepiece.

In order to focus an eyepiece, the telescope should be pointed at a light background and the eyepiece turned until the graticule lines are in sharp focus. By rotating the knurled knob on the telescope body, the focusing lens can be moved to a position when the object is in sharp focus and no parallax observable from relative movements between the graticule and image of the object being viewed.

8.2.1
Hilger and Watts telescope for optical tooling with 25 × magnification at infinity and focusing range down to 25 mm. Alignment of optical and mechanical axes is to within 3 seconds of arc at infinity focusing. The graticule is also shown.

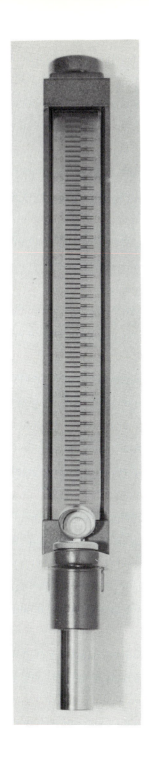

The object could be a mounted sighting scale for measuring displacements relative to the optical axis of the telescope (Fig. 8.2.2).

If the telescope is required as an autocollimator, then a Gauss type of eyepiece must be substituted for the standard eyepiece, which incorporates a filament lamp and a partial reflector, to illuminate a graticule in the eyepiece. When the telescope is focused to infinity, an image of the illuminated graticule is projected through the objective lens.

To 'square on' a part, or a surface to the line of sight, a reflector is attached to the part or surface so that the projected image is reflected back into the telescope. The return image is viewed through an eyepiece and 'squareness' achieved when the return image is superimposed on a graticule (Fig. 8.2.3)

8.2.2
Mounted sighting scale for measuring vertical displacements relative to the optical axis of the telescope. Four different widths of graduation line suit sightings from various distances.

8.2.3
Autoreflection mirror provided with cross-lines, spirit level in the top of the mount for setting the target vertical and fitted with spigot-fixing mount.

8.3 Microalignment telescopes

The telescope line of sight is a basic reference for all optical measurements and can be used for checking parallelism or squareness, and also plane and horizontal surfaces. Setting up aircraft assembly jigs, shipbuilding, ship repair, locomotive engineering, checking main crankcase bearings of large diesel engines, checking bore straightness of plastics extrusion machines, measuring displacements of turbo-generator blocks, checking turbine casings and the erection of rolling mills all contain large-scale measurement problems which have been solved with the use of microalignment telescopes.

The general arrangement of a Taylor–Hobson telescope consists of a telescopic sight built in a hardened-steel barrel with a focusing range of 25 mm to infinity. A magnification of 34× with standard eyepiece, an accuracy of ±0·05 mm at 30 m range, a line of sight coaxial with the tube which passes through the centre of the mounting sphere, and an optical micrometer block for measurements at right-angles make this instrument particularly useful for setting up aircraft assembly jigs (Fig. 8.3.1).

The exact location of the line of sight, with reference to a workpiece, must be known and so two spheres are used as references at a known height from a mounting surface. A target is mounted in one sphere with the centre of a pattern at the centre of the sphere (Fig. 8.3.2). The telescope is mounted in the other sphere and its position adjusted by pivoting it in the spherical bearing formed by the sphere and a conical mounting cup (Fig. 8.3.3).

Alignment targets parallel to within 20 seconds of arc are normally used for direct viewing, but parallelism of faces within 2 seconds is necessary if a distant target is to be viewed through an intermediate target. Most targets are calibrated and the graduations are used when the target centre is displaced more than 1·25 mm from the line of sight, which is beyond the range of the micrometers. The standard circular target, suitable for use up to about 30 m from the telescope, consists of a

centre dot and a series of concentric rings. The dimensions on the target refer to the middle of the gaps (Fig. 8.3.4).

A target illuminator for 2¼ in (57 mm)

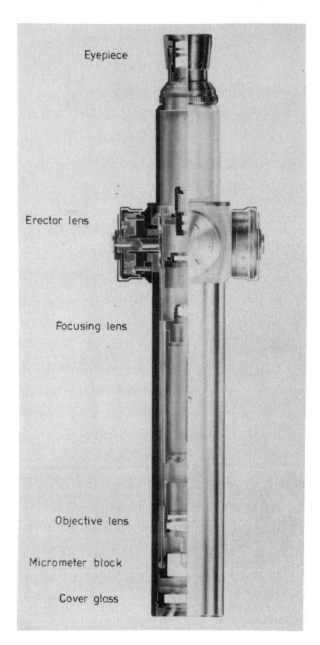

8.3.1
Taylor–Hobson microalignment telescope with 34× magnification, 600 mm field of view at 30 m range and focusing range from 25 mm to infinity.

218

8.3.2
Telescope line of sight between the centres of two spheres.

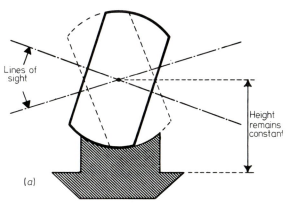

(a)

(b)

8.3.3
(a) If the line of sight passes through the centre of two spheres then, whatever the tilt of a telescope, the height above the base remains constant and so provides a datum. (b) The mounting sphere accepts either telescope or target.

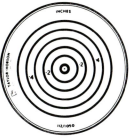

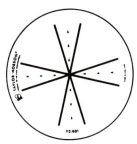

8.3.4
Alignment targets.   (a) Standard circular   (b) Cross vees   (c) Long distance

8.3.5
Target illuminator, with diffusing screen and colour filter, on mounting and flange cup.

diameter targets includes a diffusing screen and colour filter to provide the best possible background illumination for target patterns (Fig. 8.3.5).

With the combination of telescope and target, small displacements can be measured such as are needed for checking straightness. A line of sight, parallel to the nominal surface, must be established between the telescope and a target at opposite ends of the surface under test. The centre of a target is set at the same height as the centre of a telescope mounting sphere so that both ends of the line of sight are at the same height above a surface (Fig. 8.3.6).

Straightness measurements are then made along the surface between two points by placing the target at intermediate positions and adjusting the telescope micrometer until an image is central to the telescope cross-lines. Displacement of the target relative to the line of sight is read directly from the micrometer (Fig. 8.3.7).

Engine crankcase bore alignment can be rapidly checked by using the first and last bores as data points, to define ends of the line of sight. Intermediate bores are then checked relative to this reference line. A bore fixture is used to hold the telescope, with an adaptor to suit the bore diameter. Another cylindrical adaptor is needed at the last bore for a datum target (Fig. 8.3.8).

With the telescope micrometers set at zero, a tilting adjustment is made to the telescope,

8.3.6
Telescope mounted on horizontal base with sphere clamp holding sphere onto flange cup.

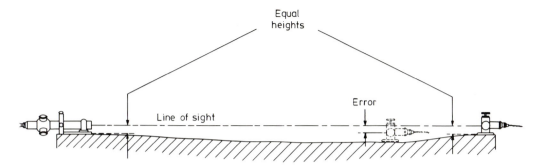

8.3.7
**Method of checking straightness by measuring
displacements from the line of sight.**

8.3.8
**Checking the alignment of bores in a diesel engine
crankcase.**

221

8.3.9
A travelling target, fitted to an adaptor plate, is used to check the misalignment of intermediate bores.

until the cross-lines are on target, to establish the line of sight. A second target and adaptor plate is then positioned in each of the intermediate bores and any misalignment measured with the telescope micrometers (Fig. 8.3.9).

Where space is cramped, or the siting position inconvenient, a right-angle eyepiece adaptor can be fitted to the microalignment telescope (Fig. 8.3.10).

The accuracy to which the stern tube of a ship can be bored depends to a large extent on the accuracy of sight off the tube and shaft bearing stands. The centre position of a hole through the after peak bulkhead is established from a target, set in a spider similar to that of the telescope, and a target is adjusted until it is centred on the line of sight (Fig. 8.3.11).

The microalignment telescope is used to provide a line of sight extending from the aft end of an unbored stern tube to a point in the engine room. From this line of sight, the tank top height, at the aft end of the engine, and the height of shaft bearing stools can be measured using a steel scale (Fig. 8.3.12).

Marine shafting can be rapidly checked for alignment by establishing a line of sight away from the shaft and then measuring the distance between this reference line and all of the shafting journals. The spheres, which are used to mount the telescope and target at the end points of the reference line, are positioned at equal heights above the journals by cups mounted on matched stands (Fig. 8.3.13).

Measurements which are offset to the line of sight, such as are required for alignment of stern tubes and shafting, are obtained by focusing the telescope onto a white and strongly graduated scale held to the bearing to be checked. The scale is first held against the bearing on which the telescope is mounted. A telescope is focused and the first

8.3.10
Right-angle eyepiece adaptor with adjustment for variations to observers' eyes from +6 to −8 dioptres. Controls for the two orthogonal micrometer movements and telescope focusing are illustrated.

**8.3.11
Spider fixture holding telescope
central in the bore of a propeller
shaft.**

**8.3.12
Line of sight from stern tube to
engine room.**

Target

Steel scales

After-peak
bulkhead

Stern post

Centre-height
of engine above
tank tops

Tank top

Stands for
shaft bearings

Target set here

Telescope
set here

223

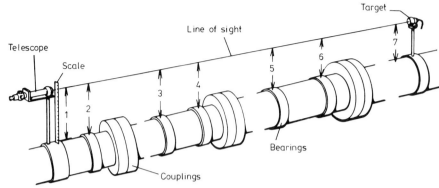

Telescope

Scale

Line of sight

Target

Bearings

Couplings

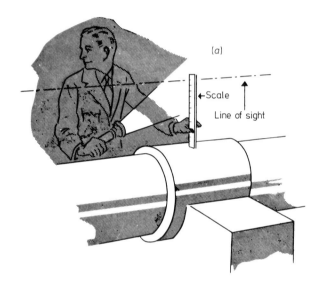

(a)

←Scale

Line of sight

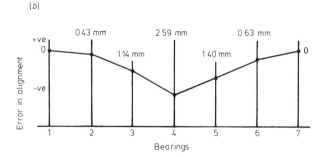

(b)

+ve

0

-ve

Error in alignment

0·43 mm

2·59 mm

0·63 mm

1·14 mm

1·40 mm

0

1   2   3   4   5   6   7

Bearings

8.3.14
(a) Steel scale used as a comparator, with initial reading treated as zero, and all measurements obtained from the telescope micrometer. (b) Example of the readings which might be obtained from marine engine shafting.

reading is treated as zero. The scale is then moved to other bearings to be checked and the readings, as measured on the telescope micrometer, will be greater or less than zero (Fig. 8.3.14).

Right-angle lines of sight are set out by 90° pentagonal prisms which have the advantage that, if the prism is correctly positioned, the angle between incident and emergent lines of sight is always a right-angle even if the prism is not set square to the line of sight. Pentagonal prisms can be fitted in a separately mounted optical square with magnetic feet, or in a sphere so that the intersection of the two lines of sight is at the centre of the sphere, or in a sweep optical square which will measure the flatness of a reference plane at 90° from the telescope axis to within ±1 second of arc (Fig. 8.3.15).

The sweep optical square is used in combination with a microalignment telescope, and three special targets, in order to establish a reference plane from which measurements can be made to a fourth target using the optical micrometer. The centre spindle of the micrometer drum is geared to the pivoting mount of a glass micrometer block and rotation of the drum results in a tilt of the glass. The target pattern is in the form of a series of bars of varying width arranged parallel with and at a constant height from the base of the target holder. The widest bar is suitable for long distances and the narrowest for short distances (Fig. 8.3.16).

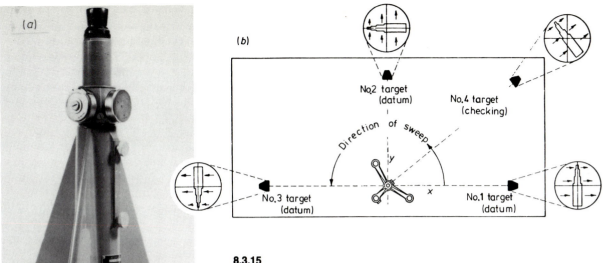

(b)

No.2 target (datum)

No.4 target (checking)

Direction of sweep

y

x

No.3 target (datum)

No.1 target (datum)

8.3.15
(a) A rotatable optical square, containing a pentagonal prism mounted on the telescope axis, sweeps out a reference plane from which errors of flatness can be measured. (b) Arrangement of sweep optical square and targets.

After a reference plane has been established, with three targets in position, a fourth target is placed where flatness is to be checked. After sighting, the optical micrometer is adjusted to bring an image central on the telescope cross-lines when the error of flatness can be read from a scale. The fourth target is then moved to other points on the surface and the reading process repeated as necessary (Fig. 8.3.17).

8.3.16
Special target for sweep optical square which is threaded to accept the target illuminator.

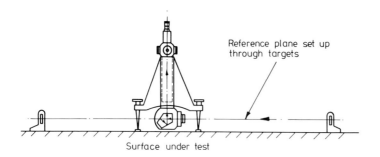

Reference plane set up through targets

Surface under test

8.4 Collimators

The collimator is a device for projecting parallel rays from a point of light and consists of an objective lens, a pinhole or graticule in the focal plane, and a light source (Fig. 8.4.1).

A silhouette of a graticule is projected through the lens and, because the graticule is at the principal focus, rays will be parallel to the line of sight as if they had come from a point at infinite distance behind the collimator.

Collimators are sometimes provided with an eyepiece, so that they can be aimed at another collimator, and may be illuminated through the eyepiece. No image is formed by beams of parallel rays but a lens placed in their path will bring each beam to a focus, on its focal plane, at a point determined by the angle between its axis and the beam of light.

Binocular telescopes are tested for alignment, and parallelism of their axes, using collimators. If magnifications produced by the two binocular tubes are not equal then an observer will experience discomfort when trying to fuse the two images and stereoscopic effects will be reduced. There may also be an error in rotation of images from the two tubes caused by misalignment of the Porro prisms.

If two tubes A and B are collimators consisting of objective lenses with graticules located at the focal plane of the lenses, and fitted with light sources, then they must be accurately aligned parallel to each other on a fixture designed to accept binoculars under test. The lens C must be located symmetrically with respect to the collimator tubes and a screen D placed in its focal plane. If axes of the two binocular tubes are parallel, then images of their graticules will superimpose on the screen. Any errors can be easily detected and measured (Fig. 8.4.2).

Collimators used for optical tooling are built in thick steel tubes, ground cylindrical on the outside to close tolerances, so that axes will be parallel to the line of sight (Fig. 8.4.3).

8.4.1

Principle of the collimator. The finite size of practical light sources results in light other than at the focus of the collimator being focused off-axis by the collimator.

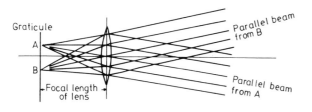

Graticule
A
B
Focal length of lens
Parallel beam from B
Parallel beam from A

8.4.2

Simplified method of testing binoculars for parallelism of axes.

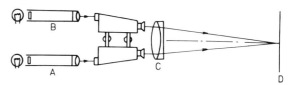

B
A
C
D

8.5 Autocollimators

A telescopic sight can be designed for use as a collimator, as well as an alignment telescope, by introducing a small light source and partially reflecting mirror in order to project silhouettes of cross-lines through the objective lens. Light is directed towards the partially reflecting mirror by a condenser lens and, when focused at infinity, the telescope becomes an autocollimator (Fig. 8.5.1).

If an optically flat mirror is attached to a part or to a surface to be aligned, and this is placed in front of a telescope, then an observer can see the reflection of his telescope in the eyepiece. The rays of light are reflected back along their own path, from a mirror set square to the line of sight, forming an image of the graticule cross-lines on the actual cross-lines themselves. If the mirror is tilted an image of the cross-lines is displaced. Because the focus is at infinity, the micrometers cannot be used for measurement but, as dark graticule cross-lines on a transparent background are used in autocollimation, the technique can be used to check squareness of the reflecting surface (Fig. 8.5.2).

General-purpose autocollimators, named Angle Dekkors or Minidekkors by Hilger and Watts Ltd, are used for measuring small angular displacements and for checking alignment, squareness, straightness and parallelism. The instrument determines angular changes to the position of a reflector relative to a datum setting from corresponding changes in the return angle of a beam of collimated light emerging from the telescope objective lens. Angular displacements in two

227

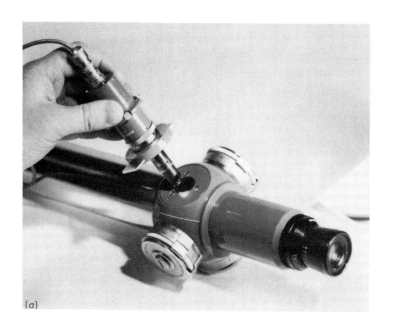

(a)

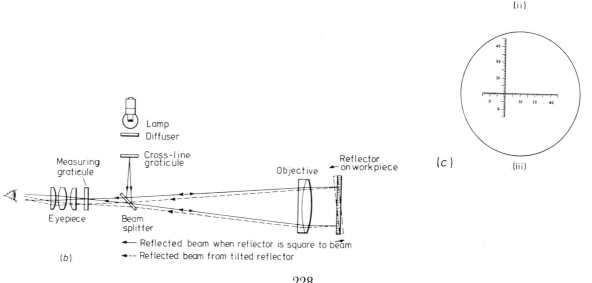

8.5.1
(*a*) **Lamphouse and partially reflecting mirror for fitting to Taylor–Hobson microalignment telescope during autocollimation. (*b*) Autocollimator optical path illustrating a telescope focused at infinity, illumination system and reflector mounted square to the beam of light. The reflector can be as small as 3 mm diameter to form a useful image. (*c*) Fixed and reflected scales.**

228

8.5.2

Autocollimators use one of the basic principles of reflection. When the reflector is tilted with respect to the incident beam, the return angle of the reflected beam is _twice_ the angular displacement of the reflector.

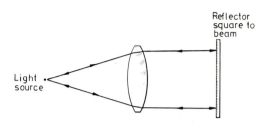

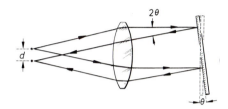

planes at right-angles can be measured simultaneously, direct from horizontal and vertical scales, through the eyepiece. Over a measuring range of 60 minutes of arc, angular displacements of 1 minute of arc can be measured and 30 seconds estimated when used as a comparator over the maximum working distance of 3·6 m. The maximum working distance at which an autocollimator can be used is limited by the aperture of the objective and the size of the returned spot of light. As the reflector is moved away from an autocollimator, the cross-line image gets shorter (Fig. 8.5.3).

The microptic autocollimator can be used to check straightness of machine tool slideways, flatness of bed plates and surface tables, and accuracy of dividing heads, because very small angles can be measured by direct reading to 0·1 second of arc. A micrometer eyepiece viewing system measures the displacement of an image from the illuminated target graticule which forms the object. An image of the illuminated target graticule in the principal focal plane of the objective lens is directed

along the axis of the telescope by a beam splitter and is brought to a focus on the eyepiece graticule. The eyepiece graticule and the reflected image of the target graticule are both viewed simultaneously through the eyepiece. No focusing adjustment is necessary because the autocollimator is always focused at infinity (Fig. 8.5.4).

A reflector is an essential part of the autocollimator and should usually be about 50 mm diameter with faces parallel to within 5 seconds and flat to 80 nm over the full aperture, but sometimes a smaller reflector (such as an angle gauge) should be used. Where reflections are required from a polished surface of low reflectivity, such as glass surfaces, the autocollimator should be fitted with a dark-field graticule and this will give an image of better visual contrast. Steel reflectors or glass Porro prisms, which give a constant deviation of 180° in one plane, are normally used for measuring departures from surface flatness (Figs 8.5.5, 8.5.6 and 8.5.7).

The Hilger and Watts photoelectric autocollimator is similar to a visual instrument but

229

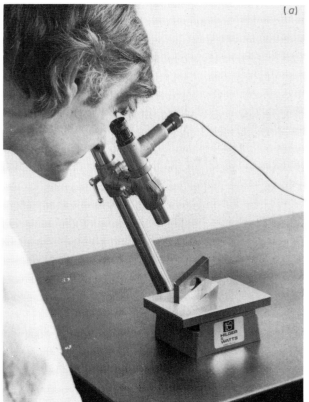

8.5.3

(a) A Minidekkor can be used for checking the angle of prisms during manufacture by comparison with an angle gauge. If reflected images of the cross-lines fall on the graticule graduations, the angular displacement can be measured. (b) Hilger and Watts angle gauges used either singly or in combination (as shown on illustration (c)) for setting up angular comparison standards. Eight angle gauges can be used for setting up any angle from 0 to 90° in 5 minute steps. The gauges are accurate to within 2 seconds of arc and flat within $\frac{1}{4}$ μm. (d) Addition and subtraction of angle gauges.

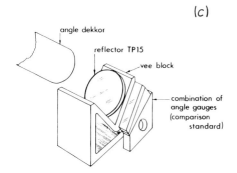

(c)

angle dekkor

reflector TP15

vee block

combination of angle gauges (comparison standard)

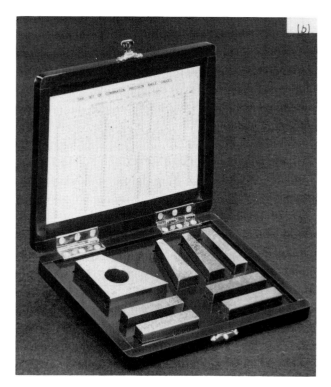

(b)

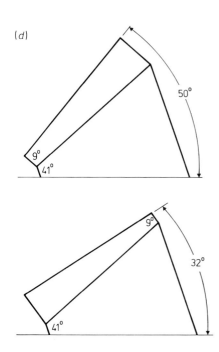

(d)

50°

9°

41°

9°

32°

41°

230

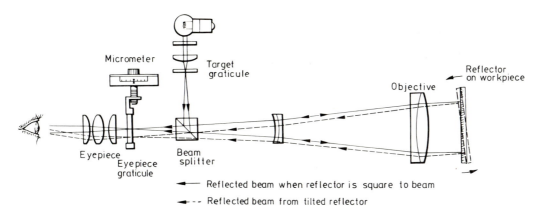

Reflected beam when reflector is square to beam

Reflected beam from tilted reflector

8.5.4
Schematic diagram of Hilger and Watts visual microptic autocollimator. Any movement which will deflect the reflector can be measured to less than 1 second of arc over a range of 10 minutes with a microptic autocollimator.

(a)

(b)

8.5.5
(a) Mounted steel reflector for vertical or horizontal viewing. (b) Mounted Porro prism for use as back-reflector. (c) Ray diagrams for Porro prism.

(c)

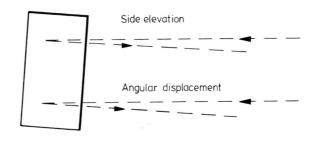

Side elevation

Angular displacement

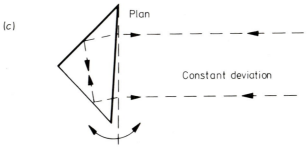

Plan

Constant deviation

231

(a)

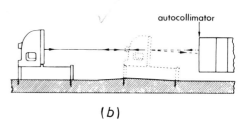

(b)

8.5.6
(a) 125 mm base reflector carriage with Porro prism giving a 1 second of arc deflection for a variation of height of 0·0006 mm.
(b) Schematic diagram illustrating use.

(a)

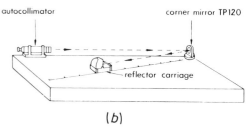

(b)

8.5.7
(a) Optically flat steel reflector for testing the angular relationship of components, steel cube with polished faces, corner mirror and 90° steel back-reflector. (b) Schematic diagram illustrating use.

with the addition of a vibrating slit and photocell assembly which detects the coincidence of eyepiece and target graticule lines. A vibrating slit assembly is connected to the micrometer screw and, when the slit is positioned so that the reflected image is within its range of movement, the intensity of light received by the photocell will vary. The output waveform is amplified and fed to a frequency discriminator and meter which indicates the asymmetry of the waveform. When the slit is positioned so that it vibrates symmetrically about the image, the meter indicates a null reading (Fig. 8.5.8.)

8.5.8
(*a*) Automatic position-sensing microptic autocollimator to monitor angular displacements in seconds of arc at a maximum working distance of 10 m. (*b*) Hilger and Watts automatic position-sensing autocollimator for measuring straightness of slideways, flatness of surface tables and calibration of polygons. Angular displacements are automatically detected and displayed on a meter reading to 1 second of arc with estimation to 0·1 second.

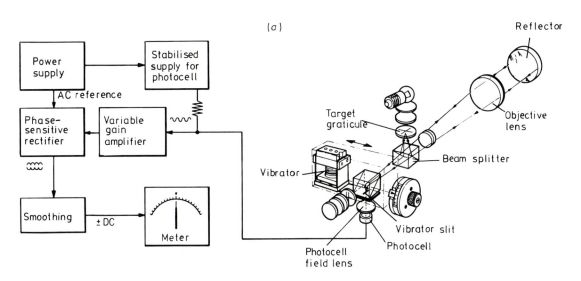

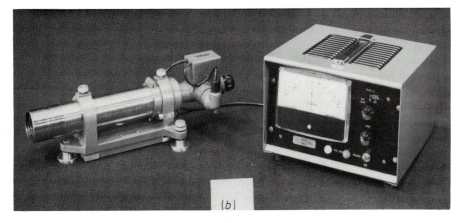

8.6 Levels and clinometers

A bubble level is one of the most sensitive optical devices and is used in most precise measuring instruments. Spirit levels depend for their accuracy on the fact that the entire free surface of a still liquid is at right-angles to the direction of gravity (Fig. 8.6.1).

A spirit level consists of a barrel-shaped curved glass tube nearly filled with alcohol, ether, petrol, methylated spirit or other hydrocarbon, according to the required viscosity and allowing for changes in temperature. The tube has a graduated scale on its upper surface and is retained in a metal casing by plaster of Paris (Fig. 6.3.1).

The radius of curvature of a bubble vial varies according to the required sensitivity. A radius of 0·5 m will be suitable for a bricklayer's level, but for a theodolite the vial radius should be at least 20 m, and for a precision block level about 150 m. A circular vial is often used for approximate levelling

and the upper surface is made as a spherical lens on which a small circle is engraved so that indications are given of level in all directions (Fig. 8.6.2).

The precision block level provides an engineer with a means for speedy levelling in assembly and installation of machine tools. Levelling is the process of determining the difference in height between two positions in the vertical or direction of gravity. A precision block level with a 10 in (254 mm) base must be tilted 10 seconds to give a gradient of 0·0006 in/ft or 0·05 mm m^{-1} or 1 in 20 000. If each division on the scale reads to 10 seconds, then by estimating one-fifth of a division on the tube a deviation from the horizontal by 1 in 100 000 may be easily checked (Fig. 8.6.3).

For most engineering work, a sensitivity of 10 seconds per division on the bubble scale can be used to measure variations of 1 to 2 seconds of arc. The bubble vial must be kinematically supported in the block level and provided with a key adjustment for calibration. Adjustable block levels are designed to measure small tilts from the horizontal beyond the range of the normal bubble scale. This is of use for testing flat surfaces which are not necessarily horizontal and for checking shaft sag. Usually one revolution of the micrometer screw will equal 1° and each division on the drum will represent 1 minute of arc.

The pendulum-type clinometer combines simplicity of operation with robustness and ease of reading. The clinometer can be used at any angle from 0 to 180° either way round and the drum can be locked so that the instrument may be handled freely without disturbing the setting (Fig. 8.6.4).

Precision microptic clinometers have divided glass scales to ensure accuracy and the horizontal plane is indicated by a spirit level. Clinometers are used for checking angular faces, gauges, and relief angles on large cutting tools and milling cutter inserts, locations of jigs and fixtures, and also levels of machine ways.

8.6.1

Large and small T bubbles reading to 8 minutes of arc, gyro bubbles reading to 6 minutes of arc per division and an adjustable level reading to 60 seconds of arc per division with a range of ±1 minute of arc.

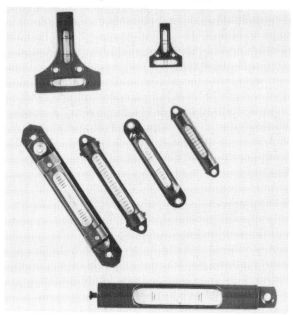

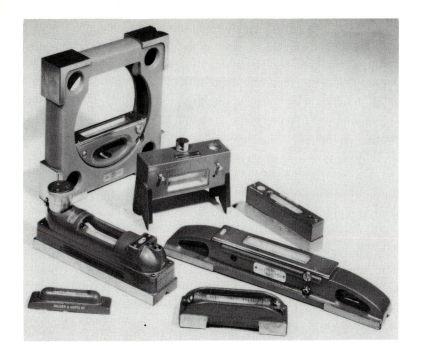

8.6.2
Square block level, stride level, angle gauge block level, precision block levels and adjustable block level.

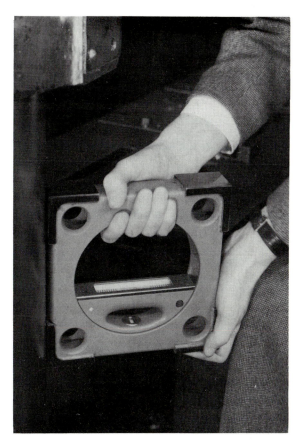

8.6.3
203 mm (8 in) square block level for checking flat or cylindrical surfaces and for setting the verticality of uprights.

235

8.6.4
Pendulum-type clinometer combining simplicity with ease of reading to minutes of arc with a fixed Vernier scale through a magnifier. The 18 in (457 mm) straight edge is 24 in (609 mm) overall length.

8.6.5
(a) Precision microptic clinometer reading to 10 seconds of arc from an enclosed glass scale which has an external spirit level mounted on it. When mounted on a special base, the instrument can be used in a horizontal position with a circular table replacing the spirit.
(b) Clinometer fitted with circular work table.

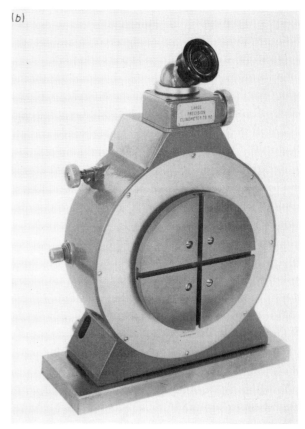

236

The microptic clinometer consists of a base enclosing a circular glass scale divided from 0 to 360° at 10 minute intervals and figured every degree. The circular scale rotates with a spirit level about a horizontal axis. Subdivision of the 10 minute graduations is made by an optical micrometer, the scale of which is divided at 10 second intervals and figured at each minute. The eyepiece of a micrometer is inclined at an angle of 45° and can be rotated to a position convenient for reading.

The spirit level has a sensitivity of 20 seconds per division with estimation possible to one-fifth of a division (Fig. 8.6.5).

A circular worktable is interchangeable with the spirit level to enable the clinometer to be used as a circular table. Alternatively, an adjustable reflector can be fitted and so enable the clinometer to be used in conjunction with an autocollimator for setting out angles (Fig. 8.6.6).

8.6.6
Clinometer fitted with adjustable mirror so that it can be used in conjunction with an autocollimator. Angle gauges are used as reference standards preliminary to checking a clinometer.

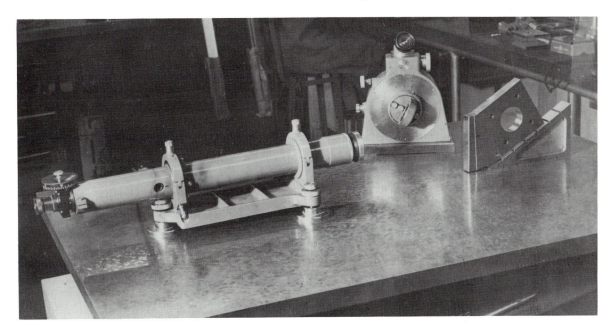

8.7 Profile projectors

Optical profile projectors are used in almost every industry for inspection and measurement of regular of irregular components. Small components or larger components, thread forms and profiles can be projected to an enlarged image on a screen where dimensions can be measured by a scale or checked against layouts and drawings.

The main advantage of optical measurement is not in the accuracy achieved but in the speed at which a test can be carried out without contact with the component. This ensures freedom from wear and no ultimate risk of false readings.

A profile projector consists of a fixed or rotatable ground-glass screen for the projected images, an optical system to form the images, a movable object table for the workpiece, and a light source and condenser lenses for diascopic or episcopic illumination.

A projection lens is used to produce an enlarged image of the object and this can be measured by the graticule of a microscope. Alternatively, the image on a ground-glass screen can be measured directly with a glass scale. In favourable conditions, the image on a ground-glass screen can be evaluated to an accuracy of 0·1 mm and this dimension governs the requirements of the optical system. Aberrations must be small enough so that the measuring error in the image does not exceed ±0·1 mm, which means that the resolving power of the system must be at least 5 lines/mm.

The image of the test object is visible on the ground-glass screen and the outer polished surface serves as a lay-on face for drawings or charts. Ordinary tracing paper is not dimensionally stable enough as it reacts to variations in humidity, and plastic foils are easier to handle but an appreciable error in measurement can occur if the room temperature varies by more than ±2 °C from the standard 20°C. If the accuracy of the projector is to be fully utilised, then errors in the reference drawings must not exceed ±0·05 mm, which can only be achieved with the aid of a precision coordinatograph.

Henri Hauser Ltd, Bienne, are leading manufacturers of optical checking and measuring instruments and their H601 profile projector has a 600 mm diameter screen with angle-measuring device, rotary objective ring and condenser turret. Diascopic and episcopic illumination is provided with standard magnifications from the lenses of 10×, 20×, 50× and 100× (Fig. 8.7.1).

The P750 projector has a relay lens system which produces an image of the object 75 mm diameter to a scale 1:1 corresponding to the maximum size of the object. This is magnified by one of the four 10× to 50× projection objectives (Fig. 8.7.2).

The objective lens produces a geometrically defined image, and all rays leaving a given point, then passing through the objective aperture, meet at a corresponding point in the image. As the ray through the centre of the objective is not deflected, the scale of image formation, or magnification factor, is determined by the ratio of the image distance to the object distance.

In the case of a profile projector, the main rays on the object side must run parallel, for measuring purposes, and magnification on the screen is independent of the position of the object. Because only the main rays are parallel, and not the image-forming rays, the word 'parallel' is avoided and instead the term 'telecentric' is used. Measurement on a fixed screen of the telecentric path of rays is independent of the position of the object and, if the test piece is moved along the optical axis, the image may lose sharpness but does not change in size (Fig. 8.7.3).

The required telecentric path of rays can be obtained by placing an aperture diaphragm in the rear focal plane of the objective to let only the telecentric beams through to produce an image of the required quality and image field. The illumination must be arranged so that it does deliver these beams of light (Fig. 8.7.4).

Ordinary projection objectives are normally provided with a central diaphragm and are not generally corrected for the telecentric path of rays. If a projector is used only for the measurement of profiles by transmitted light, it does not matter if the telecentric path of rays is produced by an aperture diaphragm in the objective of a telecentric form of illumination. A difference arises only with episcopic illumination (Fig. 8.7.5).

Projectors are usually arranged for diascopic lighting and will only produce images of transparent slides or films. Workshop measuring instruments, such as profile projectors or measuring microscopes, produce only shadow images so only the contours of test objects can be measured.

This is a restriction, as sometimes the surface of the test object should be illuminated to get images capable of evaluation, but this

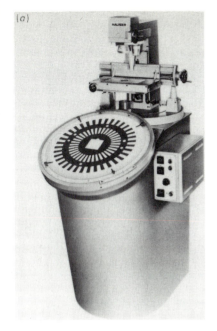

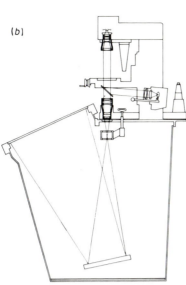

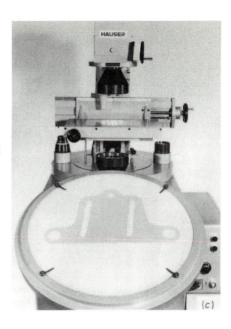

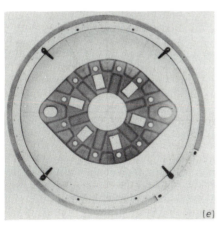

8.7.1
(*a*) Hauser H601 profile projector. (*b*) Section through Hauser H601 profile projector. (*c*) Hauser H601 profile projector equipped with large object table (410 mm × 210 mm) and objective for magnification 5×. Field of vision on object with objective 5×: diameter 120 mm. (*d*) Jewelled pallet fork. Magnification: 100×; diascopy. Checking position of fork and jewels. (*e*) Porcelain insulator. Magnification: 10×; diascopy and episcopy simultaneously. Checking the surface and the positions of the holes. (*f*) Valve. Magnification: 10×; diascopy. Checking the angle.

239

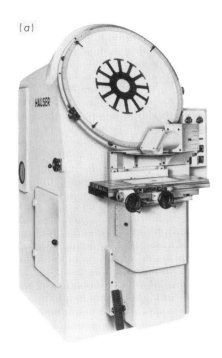

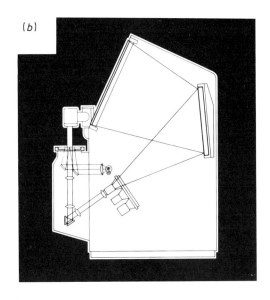

8.7.2
(a) Hauser Type P750 profile projector. (b) Schematic diagram illustrating use.

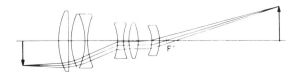

8.7.3
A special objective calculated for a telecentric path of rays.

8.7.4
A telecentric condenser.

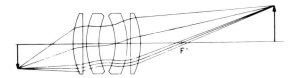

8.7.5
An ordinary projection lens used with a telecentric diaphragm at F′ or with only a telecentric condenser such as that illustrated in Fig. 8.7.4.

then eliminates one of the essentials for reliable and accurate measurement. The illuminated surface acts as a secondary source, which diffuses the incident light, and the telecentric illumination desirable for accurate measurement is not practicable.

For this reason, episcopic illumination is only used as an expedient and the quality of the image obtained is dependent on the surface structure of the object and the illumination must be adapted to the surface quality. As only the incident light thrown back into the objective can be utilised for image formation, the images are usually not very bright (Figs. 8.7.6 and 8.7.7).

Because the space between the objective lens and the object is filled by the image-forming rays, it should be kept as free as possible from interfering structural parts, so light can only be directed onto the object at an angle. Lamps may be arranged on the side of an objective, or in a ring around it, or alternatively light can be directed by movable mirrors. If the object has a highly polished surface, the oblique incident light is reflected by the object at the angle of incidence and does not enter the objective (Fig. 8.7.8).

Only if the surface diffuses the light does any appreciable part of the reflected light enter the objective. The episcopic image of the surface will look different depending on whether the radiated oblique light runs transversely or longitudinally to the grinding or turning marks of a machined surface (Fig. 8.7.9).

8.7.7
Episcopic illumination with vertical incident light via a partially transparent mirror.

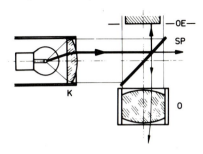

8.7.8
Oblique illumination of a highly polished surface.

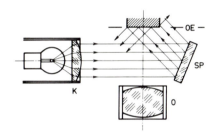

8.7.6
Episcopic illumination with oblique incident light.

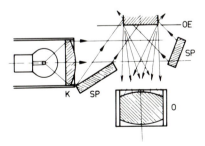

8.7.9
Oblique illumination of a diffusely scattering surface.

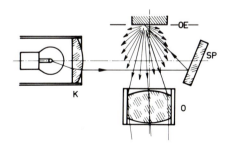

241

8.8 Interferometers

Newton's rings

Sir Isaac Newton (*Opticks*, Book II) placed a shallow-curve convex lens in contact with a plane glass plate and found that the variation in thickness of the air film, from the centre point of contact, produced concentric alternate light and dark rings when illuminated with monochromatic light. The thickness of the film of air at any point of such a system of rings or bands can be determined by counting the number of rings from the point of contact, and for practical purposes the thickness change from one ring to another is one half of the wavelength of the incident light (Fig. 8.8.1).

It is standard practice in the optical industry to compare polished lenses and flats during manufacture with test plates under helium light, and so rings or bands will be separated by $0 \cdot 28 \ \mu m$ representing one half-wavelength. If both surfaces are spherical and of equal curvature, they will show a uniform colour because the air film will be of uniform thickness (Fig. 8.8.2).

Instead of helium light, some optical shops use low-pressure mercury-vapour light, and others use daylight when the fringes will be coloured. Parallel-sided optical flats provide a convenient method of checking the faces of micrometers and other measuring instruments. Both surfaces in contact must be very clean, as a stray particle of dust can destroy reliable measurement and possibly prevent the formation of Newton's rings. It is most important that when viewing the interference pattern a nearly perpendicular line of sight to the surface is observed (Fig. 8.8.3).

Fizeau's interferometer

In 1862, A. H. L. Fizeau (1819–96) developed a non-contacting comparator in which fringes are seen from the interference of light waves between a test plate surface and the optical element under examination. The advantage of this system is that the test can be carried out without risk of damage to polished surfaces. In addition, flats can be tested up to 300 mm or more in diameter, with a high degree of accuracy, and this would be difficult to achieve by any other method.

Each incident ray of light is split into two

8.8.1

(a) Observation of Newton's rings or bands. (b) Newton's bands spaced at a half-wavelength.

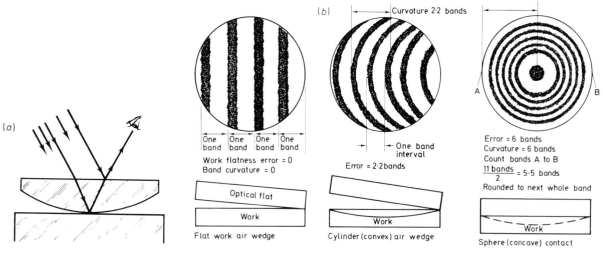

242

8.8.2
Monochromatic light sources for testing
lenses and flat surfaces by means of
Newton's rings or bands.

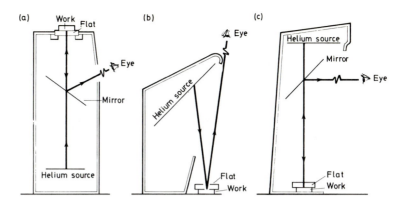

parts when it reaches the lower face of the optical flat. One portion is reflected back and the other is transmitted to the optical surface under test where it also is reflected and may be combined with the first portion in the eye. The phase relationship between the two rays of light depends on the path difference, and the phase reversal of one half-wavelength which takes place when light is reflected in air at the surface of a denser medium. If the difference in path length of the ray reflected

from the lower face of the optical flat and the ray transmitted to the reflecting surface is a whole number of wavelengths then, taking account of the phase reversal when they are recombined, they will cancel one another to give an area of darkness and so form a dark fringe (Fig. 8.8.4).

A 300 mm diameter collimating system with silica reference flat can be used for the routine inspection of blocks of prisms and flats to an accuracy of one-tenth of a fringe. The light

8.8.3
When inspecting glass test plates in contact with lenses or flats, it is important that the angle for viewing is as near normal as possible.

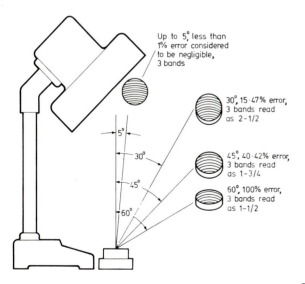

8.8.4
(a) Formation of interference fringes on a flat surface viewed under an optical flat in a parallel beam of monochromatic light. (b) Interference fringes on a flat surface viewed under an optical flat in a parallel beam of monochromatic light.

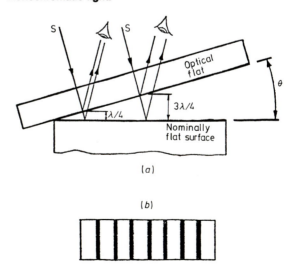

source may be a high-pressure mercury light of 546·1 nm wavelength, when the separation between the reference surface and surface under test can be only a few millimetres. If a larger spacing is necessary, then a helium–neon laser with coherent light of 632·8 nm wavelength is required for the source.

Block gauges must be finished to the highest possible accuracy, and for durability are made from stainless steel, tungsten carbide or chromium carbide. The working surfaces must be flat, so that they will adhere when wrung together, and parallel to each other to ensure equality of size at all positions between the faces. Because of the flatness and reflectivity of block gauges, the principles of optical interference, with a finely divided scale of natural origin, can be used for inspection and measurement of defects (Fig. 8.8.5).

Instead of broad Fizeau fringes with light and dark bands of equal width, an improved definition can be obtained by coating the optical flat with a uniform thin film of bismuth oxide or silver, and so increase the reflectivity to match the polished steel gauges. Because of this highly reflecting but partially transmitting film, the light will reflect back and forth between the two surfaces to form narrow fringes with wide separations and so detect small surface irregularities (Fig. 8.8.6). The length of a slip gauge is usually defined as the distance between the mid-points on opposing faces of the gauge plus one wringing film thickness. Because three or more gauges may have to be wrung together, any errors in dimensions must be determined and calibrated to within 0·25 μm.

The N.P.L.–Hilger gauge interferometer is

8.8.5
(a) The N.P.L.–Hilger gauge interferometer for measuring batches of 18 gauges wrung onto a circular steel platen. (b) Optical system of the N.P.L.–Hilger gauge interferometer: A, discharge tubes; B, condenser lens; C, entrance slit; D, achromatic objective; E, constant-deviation prism; F, optical flat; G, gauge under test; H, platen; J, mirror.

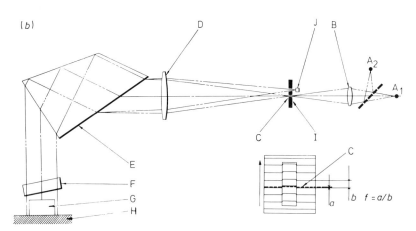

244

designed for the routine measurement of quantities of gauges up to 100 mm in length. Surface flatness and parallelism, for comparing the lengths of nominally equal gauges, can be checked and the difference in length between two gauges can be measured using three or four known light waves together with a slide rule and tables. The interferometer has two sources of illumination: a mercury-198 electrodeless lamp will provide green, blue and violet; also, a cadmium lamp will give red light of good spectral quality (Fig. 8.8.7).

A circular steel platen or base plate 150 mm diameter will take a batch of 18 gauges wrung in radial formation leaving small gaps in between for viewing fringes on the base plate. A spare base plate, loaded with a batch of gauges, should be kept inside the thermally insulated case of the interferometer ready for mounting and continuous testing.

Two sets of parallel interference fringes will be seen, one set from the face of the gauge and another from the surface of the platen. In order to measure the length of a guage, it is necessary to determine the fringe fraction for three or four wavelengths. Because barometric pressure affects the refractive index of air, and therefore the wavelength of light, accurate

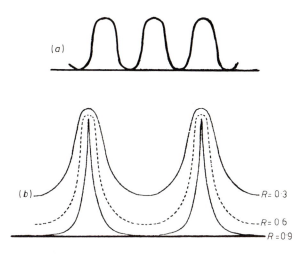

8.8.6
(*a*) Fizeau fringes are formed when two equal beams of light interfere and follow a curve in which the light and dark areas are equal. The fringes are contour lines in which the height interval between the contours is half a wavelength. (*b*) The intensity distribution varies with reflectivity for multiple-beam fringes given by a plane-parallel plate. The fringe width and the value of the minimum between the fringes characterise each distribution.

8.8.7
Interpretation of fringes as seen from the interferometer eyepiece on the gauge under test and the surface of the platen.

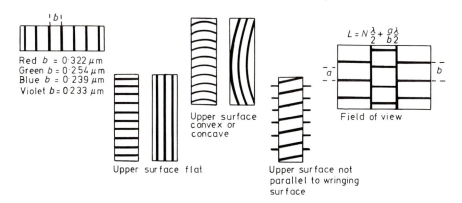

Red $b = 0.322\,\mu m$
Green $b = 0.254\,\mu m$
Blue $b = 0.239\,\mu m$
Violet $b = 0.233\,\mu m$

Upper surface flat

Upper surface convex or concave

Upper surface not parallel to wringing surface

$L = N\frac{\lambda}{2} + \frac{a}{b}\frac{\lambda}{2}$

Field of view

245

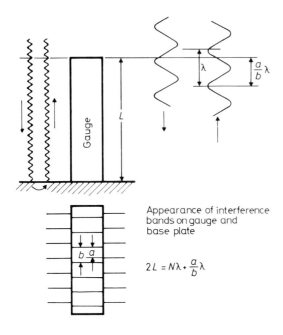

Appearance of interference
bands on gauge and
base plate

$$2L = N\lambda + \frac{a}{b}\lambda$$

8.8.8
Measuring slip gauges with a scale of light waves over twice the length of the gauge.

results can only be achieved if corrections are made for ambient pressure and temperature (Fig. 8.8.8).

When a gauge interferometer is used for measurement, twice the length of the gauge is measured with very fine scales consisting of continuous trains of different coloured light. The 'measuring scales', formed by trains of light waves, do not have any graduation lines. So, from observation of the interference bands formed on the upper face of the gauge and on the surface on which the gauge is wrung, one estimates the residual fraction of the wavelength in twice the length of the gauge for three or four different light waves (see section 6.4).

The estimated fractions, and a knowledge of the nominal size of the gauge, enable one to deduce accurately the number of whole wavelengths of each colour in the double length of the scale (see *Optical production technology* (section 11.8) and also *Dividing, ruling and maskmaking* (section 7.2)).

9 Spectrochemical analysis

9.1 Introduction

Light of different colours is refracted at different angles when passing obliquely from one transparent isotropic medium into another. A ray of white light is dispersed by a glass prism into its constituent colours because transparent materials have a different refractive index for each wavelength (Fig. 9.1.1).

This phenomenon was investigated by Sir Isaac Newton (1642–1727) in 1672, and he demonstrated that white light which had been dispersed by a prism could be reconstituted into white light by passing the coloured light through another similar prism (Fig. 9.1.2).

The arrangement of prisms used in Newton's experiment does not disperse a pure spectrum because, if the beam is broad enough to provide sufficient light, the colours overlap to form an impure spectrum. In order to produce a pure spectrum, a collimated beam of white light must be passed through a prism and then focused by a telescope to separate and image each colour.

The essential features of a simple prismatic spectroscope are that light from a source enters a narrow slit, placed at the focus of an achromatic lens, in order to produce collimated light. The light is refracted by a 60° dispersion prism which separates the wavelengths, and is collected by another lens which forms an image of the original slit at

9.1.2
Recombination of colours by an inverted prism–Newton's experiment. The emergent white ray is tinged blue on one side and red on the other.

9.1.1
Passage of white light through a prism.

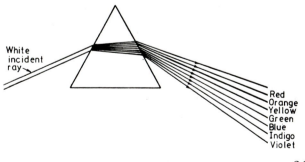

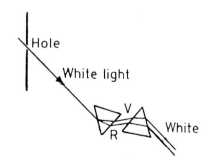

247

various points depending on the wavelength. If the source is luminous gas or vapour, produced by an arc or spark, then the emission line spectra can be observed through an eyepiece and recorded photographically or photoelectrically by means of photomultipliers (Fig. 9.1.3).

Joseph Fraunhofer (1787–1826) is famous for his development of telescope objectives, and improvements in glass melting technology, based on scientific lens design and a knowledge of the dispersion properties of optical glasses which he accurately measured in 1813.

Fraunhofer wanted to determine the dispersion of different coloured rays in order to find out whether the refracting medium for solar light was the same as for artificial light. In the course of these experiments, he found in the spectrum of sunlight a large number of strong and weak vertical lines on the spectrograph and so discovered about 574 missing wavelengths or absorption lines in the visible region of the solar spectrum.

Certain elements in the gases surrounding the Sun absorb the energy they need from the Sun's rays, in order to make them excited, and so make gaps in its spectrum. By examining the spectra of the individual elements, it is possible to discover the elements responsible for the gaps. The absorption of some wavelengths, and the resulting absorption spectrum, provides a definite means of identification of the elements.

The mechanism concerning the formation of Fraunhofer lines was explained by Kirchhoff and Bunsen who are credited with founding qualitative spectrochemical analysis. Kirchhoff's Law (1859) states that the relation between the powers of emission and powers of absorption for rays of the same wavelength is constant for all bodies at the same temperature. The first quantitative method of analysis by the spectroscope was patented by Lockyer in 1873.

The basis of spectroscopy is that every atom and molecule has a unique and characteristic relationship to light. All incandescent gases emit a discontinuous spectra of narrow bright lines, and, if that same gas at room temperature has a multispectral light passed through it, then it will absorb exactly the same wavelengths as those of its emission spectrum. If the same gas is illuminated with a high-intensity monochromatic beam of light, it will absorb and then emit light at a different wavelength to the incident light.

The physical and chemical properties of radiating bodies can be determined by their emission or absorption spectra. Many non-luminous solids, liquids and gases can be analysed by placing them between a continuous light source and entrance slit of the collimator of a spectrometer. Continuous spectra are characteristic of incandescent liquids and solids such as lamp filaments.

Atomic absorption is similar to molecular absorption, but free atoms are substituted for molecules, and electron energy levels of the atoms substituted for the vibrational and rotational levels of the molecules. The first commercial spectrograph for analytical purposes was designed in 1908 by Frank Twyman (1876–1959) of Adam Hilger Ltd.

In order to observe a spectrum, a source of light is needed, which may be an arc, spark or flame, as well as a dispersing instrument and a detector of the light such as an eye, a photographic plate, or photomultiplier and multichannel spectrometer. If the detector is a photographic plate, then a microdensitometer is needed to measure the degree of blackening of the lines. If a photomultiplier is used, then electronic equipment must be available to measure the amount of electrical energy resulting from the intensity of the spectral lines.

For metal analysis by emission spectroscopy, the source is usually an arc or spark. The sample is one electrode opposing a

9.1.3
Method of producing a pure spectrum.

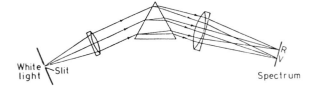

White light — Slit

Spectrum

248

counterelectrode of some material which will not confuse the analysis, and occasionally two electrodes of the sample material are used.

A source unit is needed to energise the discharge, and this may be a simple condensed spark, a triggered condenser discharge, a direct-current arc or an intermittent alternating-current arc.

The plateholder of an emission spectrograph is mounted in such a way that a number of spectra may be exposed on one plate in succession in order to divide the time of processing the plate among a number of samples. The larger the spectrograph, the more widely are the spectrum lines separated on the plate, but a longer exposure time is needed (Fig. 9.1.4).

After processing the plate, spectra will be seen to consist of a large number of lines varying in blackness. The lines due to particular elements always lie in the same places and may be observed through a spectrum projector, a comparator or a microphotometer. The microphotometer can be used to measure blackening of a line, which is a measure of the abundance of an element in the sample. A limitation is that photographic spectrographs are not suitable for measuring element concentrations greater than 10%, whereas direct-reading spectrometers do not have this limitation.

If photomultipliers are used for detection, then exit slits are mounted in position so that they will select the spectral lines of chosen elements and mask the intermediate regions. Mirrors behind the slits direct the rays onto the multiplier photocells. When light of a selected wavelength falls on its photocell, a current is excited which charges a condenser and, after an exposure, the charges on individual condensers are measured and displayed on a meter or counting system (Fig. 9.1.5).

Steel and cast iron are analysed by their emission spectra and, because the elements essential to steelmaking have useful spectral lines in the ultraviolet (air is opaque to these wavelengths), it is necessary that the spectrometer is used in a vacuum. Below 200 nm wavelength, glass and quartz will not transmit, so the prisms, lenses and windows are made from fluorite. Alternatively, the prism is replaced by a diffraction grating which reduces stray light and gives good dispersion throughout the entire spectrum (Fig. 9.1.6).

Simple compounds can absorb or emit electromagnetic waves over the whole spectrum, from microwaves to ultraviolet and x-rays, but we are only interested in that portion of the spectrum which will provide the information for a particular analysis. The spectra can therefore be divided into broad sub-divisions, such as emission, absorption,

9.1.4
From the top downwards, the spectra are those of cadmium, spelter and zinc. Note that all the zinc lines, but only the strongest cadmium lines, are reproduced in the spectrum of spelter.

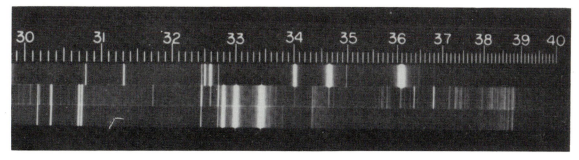

249

9.1.5
Vacuum spectrometer with fluorite prisms for the analysis of steels.

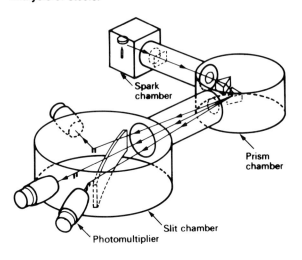

9.1.6
Diffraction grating alternative to prism for diffracting light to form a spectrum and focus it through a lens onto a sensor.

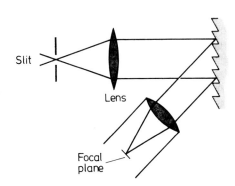

fluorescence, ultraviolet and x-ray fluorescence.

Most analyses are carried out by comparing the spectra of a sample with that of a standard, so a range of standard samples are required, containing the elements of interest, in known concentrations above and below the required concentration.

In the spectra of standard samples, the line intensities are compared with those of a line of the base element, such as iron in the case of steel. Alternatively, if no element present can be regarded as a base element, an element unlikely to be already present is added in a fixed proportion by weight and treated as a base element. The ratio of intensities of the element line and base line, plotted against a concentration of the element, constitutes a working curve and analyses of unknown samples are derived by interpolation.

Whatever the method used for analysis, the accuracy is unknown, but the result should be within a certain margin of error, determined by statistics, known as the standard deviation. This quantity is determined by finding the average of the concentrations measured and deviations from the average of those measurements.

When carrying out an analysis, there is usually time for only one determination and this result may be accepted within a degree of confidence, probability, or chance.

The fundamental advantage of a spectrograph over the direct-reading spectrometer, using photomultipliers and automatic print-out, is that it has the ability to record simultaneously tens of thousands of spectral lines at one exposure covering all elements present in the sample.

On the other hand, not all of the spectral elements may be of interest and, if the spectrum is not unknown, the linearity of response, the immediate recording of the reading and very high sensitivity for a single spectral element make the photomultiplier and multichannel spectrometer very attractive for routine industrial analysis. The main difference between the two methods of recording is in the equipment cost and the speed and accuracy with which results are obtained.

250

There are many different types of spectrometer, based on the classical design of the earliest spectroscopes consisting of a slit and dispersing system. As the basic principles are the same, only a few specific pieces of equipment are described in this chapter.

The wavelength spectrometer has a constant-deviation prism with wavelength drums in the form of calibrated helical scales on interchangeable sleeves. When fitted with a glass prism of refractive index 1·65, it will be suitable for wavelengths from 380 to 900 nm. A quartz prism will cover the range from 200 nm to 3·5 μm, a calcium fluoride prism from 380 nm to 9·0 μm, a sodium chloride prism from 380 nm to 17·0 μm, and a potassium bromide prism from 410 nm to 26·0 μm wavelength.

With suitable lamps or gas discharge tubes and accessories, this equipment can be used for measuring wavelengths, for providing monochromatic light, for absorption spectrophotometry, as well as for emission spectroscopy (Fig. 9.2.1).

There is an optimum slit width for every method of illuminating the collimator slit and this directly affects the sharpness of the spectrum lines. If the slit is widened above the critical value, the resolution is reduced. If the slit is narrowed, the intensity falls without an improvement in resolution.

The use of a spectrometer in chemical analysis depends on finding a satisfactory way of exciting the spectrum. Of the three common sources (arc, spark and flame), the source which is best suited to exciting the spectrum without damaging the sample is the spark, but it is not easily used with non-conducting powder. The spark is often used when alloys have to be analysed as it develops very little heat and low-melting-point metals can be excited without change of shape.

The arc between carbon or graphite electrodes is easily applied to inorganic powders but it does not leave the specimen unchanged. Copper rods may be used instead of carbon as the copper spectrum exhibits a number of lines, well spread out, which can be used to assist in fixing the wavelengths of unknown lines. The bunsen flame is too low in temperature, so oxycoalgas or the oxyacetylene flame is often used to analyse samples of plant and animal material.

If a liquid, such as blood, is to be examined, a small sample is placed on rolled filter paper and then the paper is burnt.

The spectrograph has two outstanding advantages over a spectroscope without direct-reading electronic equipment. One advantage is that a record can be made of ultraviolet and infrared wavelengths, and also the photographic plate is a permanent record that can be examined at leisure with the help of comparators and microphotometers.

Hilger and Watts D330 monochromators are instruments designed to isolate narrow portions of the spectrum formed by a diffraction grating. By moving the spectrum of a source internally past the slit, the wavelength of the emitted light can be varied. The plane grating is illuminated with collimated light and the quality of the spectral image depends on the configuration of the image-forming optical elements as well as on the diffraction grating (Fig. 9.2.2).

The Czerny–Turner (1930) monochromator consists of two equal-radii concave spherical

9.2.1
Wavelength spectrometer.

251

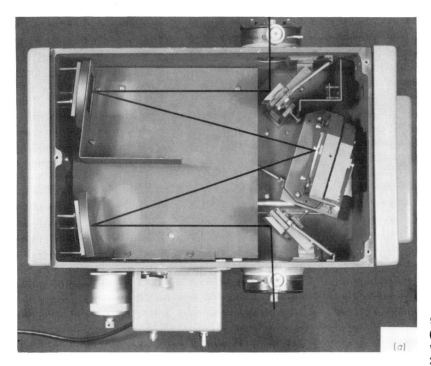

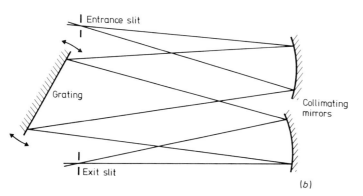

mirrors, a plane grating and curved entrance and exit slits. The first mirror collimates the light directed to the plane grating, whilst the second mirror focuses the parallel dispersed light from the grating into a spectrum. The wavelength is changed by rotating the grating on its axis from the drive system coupled to the wavelength drum.

The Ebert–Fastie (Ebert 1889, Fastie 1952, 1953, 1958) type of spectrometer differs from the Czerny–Turner in that the two separate mirrors have been combined to form one larger single mirror. A plane grating is placed at the focal distance from the mirror to direct the exit and entrance beams parallel to each other. Where complex spectra have to be analysed, such as for alkali metals in the analysis of rare earths or the determination of niobium in steel, a large Ebert spectrograph will resolve very close lines and provide dispersions on a photographic plate of less than 1 Å mm⁻¹ (Fig. 9.2.3).

Diffraction gratings consist of parallel ruled grooves which are usually straight and equally spaced on plane or concave optical surfaces. Joseph Fraunhofer developed the geometrical

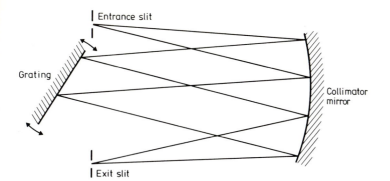

theory and ruled small gratings in 1823, in order to calibrate the wavelengths of solar absorption lines, but H. A. Rowland (1883) invented the concave grating and made gratings up to 6 in ruled width with more than 100 000 grooves (see *Dividing, ruling and mask-making*, section 4.1).

Whereas the dispersion of a prism varies with the wavelength of light, the dispersion of a grating is constant over the whole spectrum. When a fixed slit is used, as with emission spectrometry, the grating monochromator provides a constant resolving power over the entire spectrum, but stray light outside of the spectral bandwidth sometimes emerges from the exit slit. The angular dispersion is the ratio of the difference between the angles of deviation of two neighbouring wavelengths to the difference between their wavelengths (Fig. 9.2.4).

The greatest limitation with prisms is their

transparency or opacity over the wavelength range, whereas absorption problems do not arise with the use of gratings which can be manufactured with good dispersion for any spectral region. In order to enhance the spectral efficiency of a diffraction grating, the groove angle is controlled so that a maximum amount of light is dispersed into the angular region over which the grating is to be used. This is known as the 'blaze angle' because the grating will light up or 'blaze' when viewed at the correct angle.

Diffraction gratings are usually ruled in aluminium, coated on optically polished glass substrates, and then recoated with evaporated aluminium and magnesium fluoride or silicon monoxide for protection of the surface. Reflectivity enhancement for infrared wavelengths is achieved by coating with gold.

Concave gratings ruled on concave spherical mirrors can be used for dispersing, and

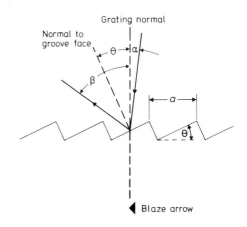

also as the entrance and exit collimator, so they are often fitted to far-ultraviolet spectrographs and direct-reading spectrometers. The 'Rowland Circle' is a circle whose diameter is equal to the radius of the concave mirror grating and, if the entrance slit is placed on the grating tangential to this circle, all monochromatic images of the slit are focused on the circle.

The laser was invented by T. H. Maiman in 1960 and this made possible what are known as holographic gratings. Two beams of coherent monochromatic light, usually from an argon-ion laser at 488 nm wavelength, are used to produce interference fringes in a photoresist which has been coated on optically polished glass. The photoresist is then developed and coated in high vacuum with aluminium or other reflecting metal. Variations in the angle of interference change the groove spacings and groove profiles can be controlled by changing the recording and processing parameters. Holographic gratings are suitable when high groove density and low noise or stray-light levels are needed.

The quantitative determination of constituents in metallurgical and chemical samples can be achieved by emission spectrography below 1000 ppm and often as low as 1 ppm for certain trace elements. Emission spectrography requires only small samples, less than 0·1 g being sufficient in most cases, and the technique can be used to solve production problems associated with the failure of materials due to contamination.

The Hilger and Watts Polyvac E900 series of direct-reading spectrometers use a holographic grating either in an air-path optical system or within a vacuum chamber. This instrument can be used to analyse liquids, powders or solids in the iron and steel industries, in the aluminium-, zinc-, copper- or lead-based industries, or in the petroleum industry. A maximum of 45 measuring channels, or any combination of them, can be used on any one analytical programme (Fig. 9.2.5).

Multichannel direct-reading spectrometers have a series of slits instead of a photographic plate, which can be set to correspond with the characteristic lines of the material to be analysed. The signal given by the photomultiplier following each slit is proportional to the intensity of a spectral line and concentration of the particular element. With the aid of a computer, many results can be printed within a minute after the specimen has been tested.

9.2.5
Polyvac system for high-performance emission direct-reading spectrometer showing light path.

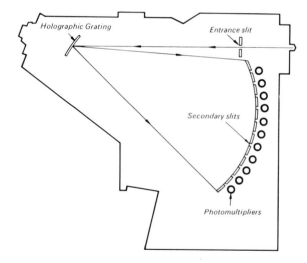

9.3 Absorption spectrophotometers

A molecular substance has a spectrum that is determined by its vibrational or rotational energy levels, and absorption spectroscopy can be carried out in the ultraviolet, visible or infrared regions. An absorption spectrophotometer measures the fraction of a range of wavelengths which have been absorbed when passing through a substance in solution.

Infrared spectrophotometers have a radiant source from which light passes through a sample and is directed onto a grating to form a characteristic spectrum. The grating is

254

scanned through the spectral range of the source and an electrical signal of varying intensity leaves the detector. Electronic equipment determines what wavelengths were absorbed in the sample, and the magnitude of the absorption, so that an analysis of the sample can be seen on a meter or drawn in chart form by a pen recorder. Infrared spectroscopy is concerned with energy imparted to or emitted by a solution under test.

Most spectrophotometers are double-beam instruments in which the difference between a sample beam and a reference beam is established. A light-chopping system alternatively gives a signal from each beam to the detector, and the two beams follow the same path except through the sample compartment. If the two beams are equal, then the detector does not provide a signal so the recorder reads 100% transmission. When a sample is inserted, differing wavelengths are absorbed, so the detector receives light of varying intensity, producing an alternating-current signal which is

used to draw absorption bands on the recorder chart.

The Hilger and Watts Infragraph can be used in the range of 2·5 to 15·4 μm wavelength for the analysis of soils and plants, organic chemicals, drugs, insecticides, paints, petroleum derivatives and polymers as liquids, solids or gases (Fig. 9.3.1).

Atomic absorption is similar to molecular absorption, but free atoms are substituted for molecules, and electron energy levels of the atoms substituted for the vibrational and rotational levels of the molecules. Free atoms are obtained by passing the sample in solution through a flame, such as air–acetylene or air–propane, and light from the source is passed through the atomised sample.

The atomic absorption technique is of particular use in the precise determination of trace metals in lead, zinc, steel, bronze, wood-pulp ash, lubricating oils, fuel oil, water, trade effluent and in a variety of other applications.

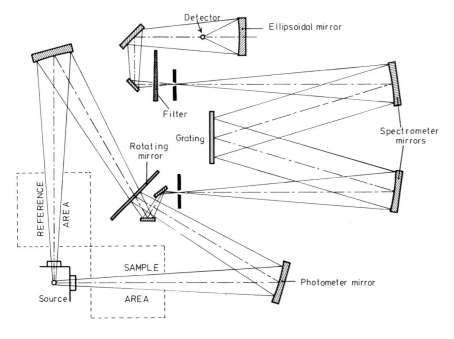

9.3.1
Optical layout of Infragraph infrared spectrophotometer.

255

10 Ophthalmic and Medical Instruments

10.1 Introduction

Spectacles were invented 700 years ago and, at that time, lenses were made from polished quartz crystal held together by wire or leather frames. The experiments of John Dollond in 1757 and Pierre Louis Guinand in 1768 led to the foundation of the optical glass industry and to the mass production of spectacles (see *Spectacle lens technology*, section 1.2).

The human eye is very similar to a camera, in that the lens must focus a picture or image of a scene onto the retina, which is connected by optic nerves to the brain. The retina consists of a mosaic of light-sensitive cells in two groups. The cones are sensitive to colour, and are the receptors for day vision, whilst the rods are insensitive to colour, and are used for night vision (Fig. 10.1.1).

In the region of the fovea, the receptors are crowded together, and consist almost entirely of cones, each of which has a separate optic nerve joining it to the brain. The density of the receptors is less in the outer regions of the retina, where they include a proportion of rods which increase in number with increasing distance from the fovea. In the outer regions, the receptors do not each have a separate nerve but are joined to the brain in groups. It has been estimated that the retina of each eye contains about 10 million cones and 120 million rods in order to provide us with high-resolution vision.

If an eye is perfectly formed, then its crystalline lens will give a sharply focused image of distant objects on the retina. When viewing near objects, the lens will alter its shape and curvature again to provide a sharp image, and this ability to focus different distances is known as accommodation.

Very few human eyes are perfect and all change their characteristics from year to year. Some people have long-sight (hypermetropia) and cannot easily focus for reading, whilst others have short-sight (myopia) and find it difficult to see clearly at a distance. As a result of age, the crystalline lens gradually hardens (presbyopia) and the power of accommodation declines, so supplementary lenses of two or more powers are needed for clear vision at all distances. Most eyes are not truly spherical so, in addition to other defects, the lens is incapable of giving a uniformly sharp image over the whole retina and this defect is known as astigmatism.

Spectacle lenses have one spherical surface to correct refractive errors and the other, cylindrical or toroidal in shape, to neutralise astigmatism. Spectacle lens prescriptions include details of spherical and cylindrical powers, axis direction, vertex distance, prism or decentration, tint and form of lens.

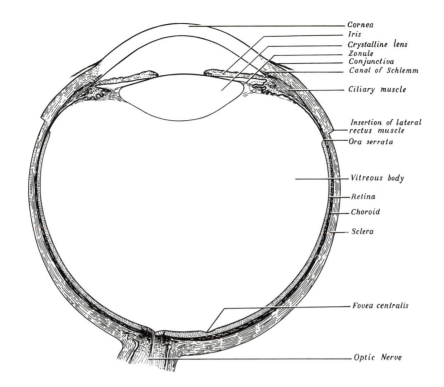

Cornea
Iris
Crystalline lens
Zonule
Conjunctiva
Canal of Schlemm

Ciliary muscle

Insertion of lateral
rectus muscle
Ora serrata

Vitreous body

Retina

Choroid

Sclera

Fovea centralis

Optic Nerve

Refraction errors can be estimated by subjective or objective methods. In 1843, Fronmüller devised a trial case of 60 pairs of plus and minus spherical lenses for the subjective assessment of lens prescriptions. Thirty years later, plano cylinders were added for the determination of astigmatism.

Since 1950, a range of ophthalmic vision testers and chart projectors, and also special surgical microscopes and cameras for retinal photography, have been developed. The invention of the laser in 1960 increased the importance of ophthalmic surgery, including the treatment of detached retinas, as well as other eye disorders, by photocoagulation.

10.2 Ophthalmic testing

An optician must determine that the eyes of a patient are healthy and then prescribe spectacles which will give the best possible vision for each eye and also, by binocular refraction, ensure that optimum vision is obtained from both eyes. By subjective refraction, the optician relies on the patient's responses to lenses placed in front of his eyes. Objective refraction means obtaining quantitative information without depending on a correct reply from the patient.

After the prescription for each eye has been determined, a modification may be necessary if it is found that, when working together, the eyes have a tendency to turn in or out and so cause visual difficulties or double vision in extreme cases.

Most refractions have depended on the patient's subjective interpretation of the quality of an image formed upon the retina using equipment containing a series of sphere dials, each with about eleven lenses of different powers, and other accessories such as prisms, cylinders and filters (Fig. 10.2.1).

The desirability of an accurate and rapid objective method of measuring the refraction of an eye has been recognised for a long time. Many patients are unable to give the correct

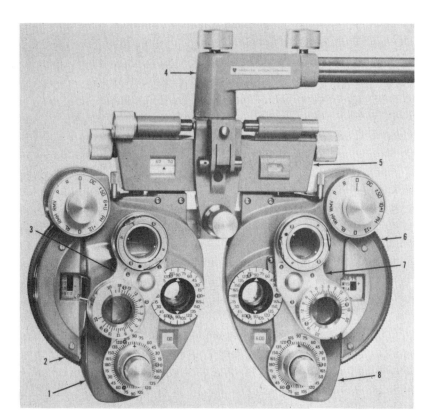

10.2.1
American Optical Ultramatic Rx
Master Phoroptor Refractor:
1, cylinder housing assembly (right
eye); 2, sphere housing assembly
(right eye); 3, turret assembly
(right eye); 4, head yoke; 5, main
support; 6, sphere housing
assembly (left eye); 7, turret
assembly (left eye); 8, cylinder
housing assembly (left eye);
sphere and auxiliary dials (not
shown above); beam and card
holder (not shown above).

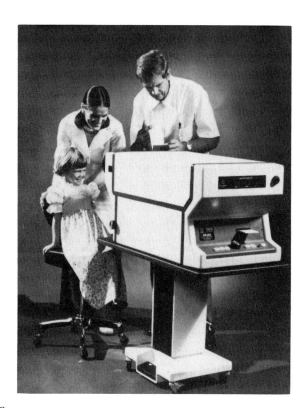

10.2.2
The Acuity Systems Auto-Refractor is designed for the
automation of objective refraction of the eye.

information which is required by the examiner. The effect of correcting lenses may at first be misinterpreted or visual acuity may be so low that alterations in the lenses are not appreciated by the patient.

In the past, many objective optometers have been produced, but they lacked the precision to measure refraction to the required accuracy of at least 0·25 dioptre and also locate the principal meridians of low amounts of astigmatism within a few degrees.

With the development of infrared diodes, emitting invisible radiation, and photo-detectors, several instrument manufacturers have made automatic objective refractors capable of prescribing for non-communicative patients, such as children, the elderly or mentally retarded, or foreigners (Fig. 10.2.2).

The Acuity Systems 6600 Auto-Refractor finds the refractive error in a patient's eye and transfers the information into a prescription for sphere, cylinder and axis.

10.3 Lens testing

The prescription supplied by an optician to a spectacle-lens manufacturer or prescription laboratory has often to be changed depending on the way in which the lenses are made. The correction for astigmatism may be applied by a cylindrical surface on the front or rear of the lens. Whatever the sign of the cylinder written on the prescription, a lens can be made with the opposite cylinder sign providing the prescription is transposed. A lens can also be made in any one of a large number of curved forms.

The rule for transposition from a plus to a minus cylinder form is to add the original sphere and cylinder together, taking signs into account, then change the sign of the cylinder and turn the cylinder axis through 90°. If 90° or less then add 90°, or if over 90° subtract 90°.

Measurement of spectacle lenses or trial case lenses is carried out with a focimeter. Vertex power, axis direction of cylinders, prism and prismatic effects, as well as lens marking are within the capability of this instrument which is essential to every optician and prescription manufacturer (Fig. 10.3.1).

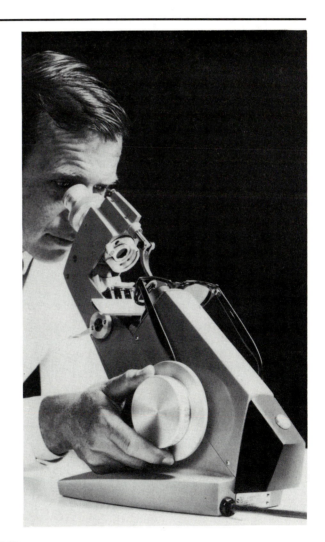

10.3.1
American Optical Lensometer for prescription lens measurement of power and axis.

259

10.3.2
Acuity Systems Auto-Lensmeter for automatic measurements of lens power and axis with digital display.

The focimeter has a focusing system and observation system. The focusing system consists of a movable target, a collimating lens and a rest for the lens under test. The observation system takes the form of a telescope and adjustable measuring eyepiece through which can be seen a rotatable graticule to determine axis direction by reference to a protractor scale.

Because accuracy of readings does depend on the operator, and the instrument can go out of adjustment, there has been a need for a reliable objective instrument. By combining the properties of light from a helium–neon laser with microprocessors, it has been possible to design an instrument capable of rapidly measuring sphere, cylinder, axis and prism either in 0·25, 0·125 or 0·01 dioptre increments. The readings are digitally displayed on a screen in standard prescription notation and can also be automatically printed onto labels or documents (Fig. 10.3.2).

10.4 Medical

Laser equipment is likely to be used for surgical and radiological treatment of malignant tumours by causing their progressive regression and eventual dissolution with only slight injury to normal tissue. Laser radiation will remove warts, birthmarks and tattoos by an operation which is fast and painless.

Ophthalmic surgeons use laser sources for photocoagulation of ocular tissues. Photocoagulation occurs when the eye changes light energy into heat and a thermal burn is produced which develops a scar as the lesion heals. Holes and tears in the retina can be repaired, and malignant or benign eye tumours can also be destroyed. The laser has made photocoagulation a relatively safe and easy operation.

In addition to special ophthalmometers, eye refractometers, slit lamp microscopes and retinal cameras for the ophthalmic surgeon, there are an increasing number of fibrescopes available for the endoscopic observation and photography of the stomach. Details concerning the manufacture and use of fibre optics are described in *Optical production technology* (section 16.2).

Appendix 1 Bibliography

The references are listed in alphabetical order of author in each chapter to which they apply.

Chapter 1

BETTS E., 1973, *Film Business: A History of the British Cinema, 1896–1971* (London: Allen & Unwin).

CLOSE, Sir Charles F. Arden-, 1978, *The Early Years of the Ordnance Survey* (Plymouth: David and Charles).

KING C. H., *The History of the Telescope*, Chapter 8 (London: Griffin).

MANVILLE G. E., *Two Fathers and Two Sons* (Newcastle upon Tyne: Grubb–Parsons).

MITCHELL M., *The Dollond and Aitchison Museum* (Yardley, Birmingham: Dolland and Aitchison).

TAYLOR W., WILSON E., SIMMS J. and MAXWELL P. D., *At the Sign of the Orrery* (York: Vickers Instruments).

THOMPSON E. H., 1935, An automatic plotting instrument for use with air photographs, *Report of Proceedings, Conference of Empire Survey Officers, 1935* (London: Her Majesty's Stationery Office).

WOOD A., *Mr Rank, A study of J. Arthur Rank and British Films* (Sevenoaks: Hodder and Stoughton).

Chapter 2

HECHT E. and ZAJAC A., 1975, *Optics* (New York: McGraw-Hill).

HOLDEN J. and PARSLEY M., 1976, Channel plates, *Engineering* July.

HORNE D. F., 1972, *Optical Production Technology* (Bristol: Adam Hilger).

JACOBS D. H., 1943, *Fundamentals of Optical Engineering* (New York: McGraw-Hill).

KING H. C., *The History of the Telescope* (London: Griffin).

MARTIN L. C., *Technical Optics* (London: Pitman).

NASA, *Optical Telescope Technology*, NASA Publication SP-233 (Washington, D.C.: U.S. Government Printing Office).

PAGE T. and PAGE L. W., *Telescopes* (London: Macmillan).

PIRENNE M. H., 1948, *Vision and the Eye* (London: Chapman and Hall).

V.E.B. CARL ZEISS JENA 1976, *Jena Review*, No. 3 (Jena: V.E.B. Carl Zeiss).

Chapter 3

BENNET, JUPNIK, OSTERBERG and RICHARDS, 1951, *Phase Microscopy* (Chichester: Wiley).

BELLING J. *The Use of the Microscope* (New York: McGraw-Hill).

CLAY R. S. and COURT T. H., 1975, *The History of the Microscope* (The Holland Press).

HARTSHORNE and STUART, *Crystals and the Polarizing Microscope* (London: Arnold).

MARTIN L. C., *Technical Optics* (London: Pitman).

MODIN H. and MODIN S., *Metallurgical Microscopy* (London: Butterworths).

TURNER G. l'E., 1971, *Micrographia Historica, The Study of the History of the Microscope*, Quekett Lecture 11 November 1971 (London: Royal Microscopical Society).

VICKERS *Optics for the microscope* (York: Vickers Instruments).

WILD HEERBRUGG, 1969, *Microskopion* **17**, September (Heerbrugg: Wild Heerbrugg).

——1973, *Microskopion* **23**, November (Heerbrugg: Wild Heerbrugg).

——1975, *Microskopion* **24**, April (Heerbrugg: Wild Heerbrugg).

ZERNIKE F., 1942, Phase contrast, a new method for microscopic observation of transparent objects, *Physica* **9**, 686–98, 974–86.

Chapter 4

ALLISTER R., 1948, *Friese-Greene* (USA: Marsland).

CAMPBELL F. W., LAW T. A., MORRIS L. F. and SINCLAIR A. T., *Sound-Film Projection* (London: Newnes).

CLULOW F. W., 1972, *Colour, Its Principles and their Applications* (New York: Fountain Press).

COOK G. H., 1973, Recent developments in television optics, *R. Television Soc. J.* Jan–Feb, 158–167.

COX A., 1974, *Photographic Optics* (New York: Focal Press).

HARRY J. E., 1974, *Industrial Lasers and Their Application* (New York: McGraw-Hill).

JAMES I. J. P., PERKINS D. G., PYKE P. J., TAYLOR E. W., KENT D. E. and FAIRBAIRN I. A., 1970, The E.M.I. four-tube colour television camera, *Radio & Electron. Engr* **39** (No. 5, May), 249–70.

MCCROBIE G. L., 1974, M.T.F. in lens design at Xerox, Image assessment and specification, *Proc. Soc. Photo-Opt. Instrum. Engr* **46** (May) 53–7.

O'SHEA D. C., CALLEN W. R. and RHODES W. T., *Introduction to Lasers and Their Applications* (Reading, Mass.: Addison-Wesley).

THOMAS D. B., *The Origins of the Motion Picture*, A Science Museum Booklet (London: Her Majesty's Stationery Office).

Chapter 5

ASHER H., 1970, *Photographic Principles and Practices* (New York: Fountain Press).

COX A., 1974, *Photographic Optics* (New York: Focal Press).

HORNE D. F., 1975, *Lens Mechanism Technology* (Bristol: Adam Hilger).

SAXE R. F. 1966, *High-Speed Photography* (New York: Focal Press).

Chapter 6

BANNISTER A. and RAYMOND S., 1977, *Surveying* (London: Pitman).

CLARK D., 1972/73, *Plane and Geodetic Surveying*, Vols I and II (London: Constable).

CLENDINNING J. and OLLIVER J. G., 1978, *Principles and Use of Surveying Instruments* (New York: Van Nostrand–Reinhold).

COOPER M. A. R., *Modern Theodolites and Levels* (London: Crosby Lockwood).

HARLEY J. B., 1975, *Ordnance Survey Maps* (Southampton: Ordnance Survey).

HORNE D. F., 1974, *Dividing, Ruling and Mask-Making* (Bristol: Adam Hilger).

ORDNANCE SURVEY, *Triangulation and Minor Control Information*, Leaflet No. 2 (Southampton: Ordnance Survey).

——*Levelling*, Leaflet No. 3 (Southampton: Ordnance Survey).

RAYNER W. H. and SCHMIDT M. O., 1969, *Fundamentals of Surveying* (New York: Van Nostrand–Reinhold).

STRASSER G. J., 1966, Heinrich Wild's contribution to the development of modern survey instruments, *Survey Review* April.

TOMALIN G., 1974, *Precision Site Surveying and Setting Out* (Bristol: Adam Hilger).

TRUTMAN O., *Levelling* (Heerbrugg: Wild Heerbrugg).

——*The Theodolite and its Application* (Heerbrugg: Wild Heerbrugg).

WARD A. H., 1968, The changing face of survey instruments *Australian Surveyor* June.

Chapter 7

BEARD L. F. H., 1976, A portable stereometric camera for clinical measurement, *Med. & Biol. Illustr.* **26**, 107–10.

BROCK G. C., *The Physical Aspects of Aerial Photography* (New York: Dover).

CIMERMAN V. J. and TOMASEGOVIC Z., 1970, *Atlas of Photogrammetric Instruments* (Amsterdam: Elsevier).

FOURCADE H. G., 1902, *On a Stereoscopic Method of Photographic Surveying*, Paper before South African Philosophical Society, October.

HART C. A. *Air Photography Applied to Surveying* (London: Longmans Green).

HOTINE M., 1927, *Stereoscopic Examination of Air Photographs*, Professional Papers of the A.S.C., No. 4 (London: Her Majesty's Stationery Office).

——*Simple Methods of Surveying from Air Photographs*, Professional Papers of the A.S.C., No. 3 (London: Her Majesty's Stationery Office).

——*Extensions of the 'Arundel' method*, Professional Paper of the A.S.C., No. 6, A continuation of Professional Paper No. 3 (London: Her Majesty's Stationery Office).

KING L. N. F. K., *Graphical Methods of Plotting from Air Photographs* (London: Her Majesty's Stationery Office).

MOFFITT F. H., *Photogrammetry* (New York: Internation).

RICHTER G., *Dictionary of Optics, Photography and Photogrammetry* (Amsterdam: Elsevier).

TALLEY B. B., *Engineering Applications of Aerial and Terrestrial Photogrammetry* (London: Pitman).

The Optical Industry and Systems Directory (Pittsfield, Mass.: Optical Publishing).

THOMPSON E. H., 1935, An automatic plotting instrument for use with air photographs, *Report of Proceedings, Conference of Empire Survey Officers, 1935*, pp 132–3 (London: Her Majesty's Stationery Office).

——1937, The seven-lens aerial camera, *R. Engr. J.* **51** 217.

——1938, The seven-lens air survey camera, *E.S.R.* **4**, (No. 26) 216; (No. 27) 263.

WINCHESTER C. and WILLS F. L., *Aerial Photography* (London: Chapman & Hall).

WOLF P. R., 1974, *Elements of Photogrammetry* (New York: McGraw-Hill).

V.E.B. CARL ZEISS JENA 1976, *Jena Review* No. 2, Surveying News.

——*Jena Review* No. 32, Photogrammetry.

Chapter 8

ANON, *Optical Alignment* (London: Rank Precision Industries Ltd, Metrology Division).

BOTTOMLEY S. C., 1967–68, Use of optical probes in metrology, Part I, *Hilger J.* **10** (No. 3); Part II, *Hilger J.* **11** (No. 1).

HORNE D. F., 1974, *Dividing, Ruling and Mask-Making* (Bristol: Adam Hilger).

JACOBS D. H., 1943, *Fundamentals of Optical Engineering* (New York: McGraw-Hill).

KISSAM P., *Optical Tooling* (New York: McGraw-Hill).

Chapter 9

CANDLER C., 1949, *Practical Spectroscopy* (London: Adam Hilger).

de GALAN L., *Analytical Spectrometry* (London: Adam Hilger).

GIRARD A. and JACQUINOT P., Principles of instrumental methods in spectroscopy, Chapter 3 of *Advanced Optical Techniques*, ed A. C. S. Van Heel (Amsterdam: North-Holland).

HEAVENS O. S., 1971, *Lasers* (London: Duckworth).

JONES E. B., *Instrument Technology*, Vol. II. *Analysis Instruments* (London: Butterworths).

KINGSLAKE R., 1965–7, *Applied Optics and Optical Engineering* (New York: Academic Press).

Chapter 10

BENNETT A. G., *Ophthalmic Lenses* (London: Hatton Press).

—— 1972, *Ophthalmic Prescription Work* (London: Butterworths).

HORNE D. F., 1978, *Spectacle Lens Technology* (Bristol: Adam Hilger).

Subject Index

Name and Author Index

Companies and Institutions Index

271

List of Companies and Institutions, with Trade Names

Acuity Systems Inc.
  11413 Isaac Newton Sq., Reston,
  Virginia 22090, U.S.A.

Trade Names
Auto-Lensmeter
Auto-Refractor

American Optical Corp.
  Scientific Instrument Division,
  Buffalo, New York 14215, U.S.A.

Fluorestar
Lensometer
Microstar
Phoropter
Ultramatic

Arnold and Richter KG
  Türkenstrasse 89, D-8000
  München 40, Germany

Arriflex

Barr and Stroud Ltd
  Caxton Street, Anniesland,
  Glasgow, G13 1HZ

Buckbee–Mears Europe GmbH
  D–7840 Müllheim, Renkenruns-
  strasse, Postfach 102, Germany

Chance–Pilkington
  Glascoed Road, St Asaph, Clwyd
  LL17 0LL

Crosfield Electronics Ltd
  766 Holloway Road, London,
  N19 3JG

Autotron
Laserdot
Lasergravure
Magnascan
Synchroscope

J. H. Dallmeyer Ltd
  Church End Works, High Road,
  Willesden, London, NW10 2DN

Trade Names
Adon
Dallon
Pentac

Dollond and Aitchison Group Ltd
  1323 Coventry Road, Yardley,
  Birmingham, B25 8LP

Ferranti Ltd
  Electronic and Display Equip-
  ment Division, Gem Mill,
  Chadderton, Oldham, Lancs

Henri Hauser Ltd
  CH-2500, Bienne, Switzerland

Hewlett-Packard
  5301 Stevens Creek Blvd, Santa
  Clara, California 95050, U.S.A.

Hunting Surveys and Consultants
Ltd
  Elstree Way, Boreham Wood,
  Herts, WD6 1SB

Kern and Co. AG
  CH-5001, Aarau, Switzerland

Mekometer

Linhof Prazisions Kamera Werke GmbH
  8 München 70, Rupert-Mayer-Strasse 45, Postfach 701229, Germany
	Trade Names
	Aero Technika
	Master Technika
		Rank Taylor Hobson
		  P.O. Box 36, Guthlaxton Street, Leicester, LE2 0SP
			Trade Names
			Angle Dekkor
			Autoplumb
			Cooke
			Microptic
			Minidekkor
			Quickset
			Talyrond
			Talysurf

Linotype UK Kingsbury Works, Kingsbury Road, London, NW9 8UT	Linotron Mergenthaler		

Littlejohn Graphic Systems Ltd
  16–24 Brewery Road, London, N7 9NP
		RCA Systems Division
		  Camden, New Jersey 08102, U.S.A.	

		Royal Observatory Edinburgh Edinburgh, EH9 3HJ	

Matra Division Optique
  93 Avenue Victor-Hugo, B.P. 209, 92505 Rueil-Malmaison Cedex, France
	Orthosfom
	Panoramat
	Traster
		United Kingdom Optical Co. Ltd
		  Bittacy Hill, London, NW7
			Univis

		Vickers Instruments Haxby Road, York, YO3 7SD	Cooke Tavistock

The Monotype Corporation Ltd
  Salfords, Redhill, Surrey, RH1 5JP
	Lasercomp
	Monophoto
		Wild Heerbrugg Ltd
		  CH-9435, Heerbrugg, Switzerland
			Autograph
			Aviogon
			Aviophot
			Avioplan
			Distomat
			Tachymat
			Tachymeter

Muirhead Data Communications Ltd Beckenham, Kent, BR3 4BE			

Pictorial Machinery Ltd
  Honeycrock Land, Salfords, Redhill, Surrey, RH1 5JP
		Carl Zeiss Oberkochen
		  7082 Oberkochen, Postfach 1369/1380, Germany
			Eldi
			Planicart
			Pleogon
			Stereocord
			Stereopret
			Telikon
			Topar
			Toparon

Rank Xerox Ltd Mitcheldean, Gloucestershire, GL17 0DD	Xerography		

Rank Hilger
  Westwood, Margate, Kent, CT9 4JL
	Fluoroprint
	Infragraph
	Polyvac
	Spekker
		V.E.B. Carl Zeiss Jena
		  DDR-69 Jena Carl-Zeiss-Strasse 1, Germany
			Spacemaster